About Island Press

Since 1984, the nonprofit Island Press has been stimulating, shaping, and communicating the ideas that are essential for solving environmental problems worldwide. With more than 800 titles in print and some 40 new releases each year, we are the nation's leading publisher on environmental issues. We identify innovative thinkers and emerging trends in the environmental field. We work with world-renowned experts and authors to develop cross-disciplinary solutions to environmental challenges.

Island Press designs and implements coordinated book publication campaigns in order to communicate our critical messages in print, in person, and online using the latest technologies, programs, and the media. Our goal: to reach targeted audiences—scientists, policymakers, environmental advocates, the media, and concerned citizens—who can and will take action to protect the plants and animals that enrich our world, the ecosystems we need to survive, the water we drink, and the air we breathe.

Island Press gratefully acknowledges the support of its work by the Agua Fund, Inc., The Margaret A. Cargill Foundation, Betsy and Jesse Fink Foundation, The William and Flora Hewlett Foundation, The Kresge Foundation, The Forrest and Frances Lattner Foundation, The Andrew W. Mellon Foundation, The Curtis and Edith Munson Foundation, The Overbrook Foundation, The David and Lucile Packard Foundation, The Summit Foundation, Trust for Architectural Easements, The Winslow Foundation, and other generous donors.

The opinions expressed in this book are those of the author(s) and do not necessarily reflect the views of our donors.

TEMPERATE AND BOREAL RAINFORESTS OF THE WORLD

TEMPERATE AND BOREAL RAINFORESTS OF THE WORLD: ECOLOGY AND CONSERVATION

Edited by

Dominick A. DellaSala

Geos Institute

Foreword by David Suzuki

ISLANDPRESS

Washington | Covelo | London

Figure 8.2 reprinted from *High-Latitude Rainforests and Associated Ecosystems of the Wet Coast of the Americas: Climate Hydrology, Ecology, and Conservation*, edited by Richard G. Lawford, Paul B. Alaback, and Eduardo Fuentes. ©1996 Springer-Verlag New York. Used with kind permission of Springer Science+Business Media.

"If a Tree Falls," written by Bruce Cockburn. ©1988 Golden Mountain Music Corp. International copyright protected. Reprinted with kind permission of the artist.

Library of Congress Cataloging-in-Publication Data

Temperate and boreal rainforests of the world : ecology and conservation / edited by Dominick A. DellaSala ; foreword by David Suzuki.
       p. cm.
  Includes bibliographical references and index.
  ISBN-13: 978-1-59726-675-8 (cloth : alk. paper)
  ISBN-10: 1-59726-675-2 (cloth : alk. paper)
  ISBN-13: 978-1-59726-676-5 (pbk. : alk. paper)
  ISBN-10: 1-59726-676-0 (pbk. : alk. paper)
  1. Rain forest ecology. 2. Taiga ecology. 3. Temperate climate. 4. Rain forest conservation. 5. Taiga conservation. I. DellaSala, Dominick A., 1956–
  QH541.5.R27T46 2011
  577.34—dc22

                                                                2010018082

Printed on recycled, acid-free paper

Manufactured in the United States of America
10 9 8 7 6 5 4 3 2 1

Book design and typesetting by: Karen Wenk
Font: Bembo

Keywords: Rainforest, temperate rainforest, boreal rainforest, perhumid coastal forest, Valdivian temperate rainforest, relict, climate change, gap dynamics, logging, road building, Knysna-Tsitsikamma forest, Colchic forest, Hyrcanic forest, humidity-dependent forest, montane forest, subpolar rainforest, seasonal rainforest

Contents

Foreword

When I attended the 1992 Earth Summit in Rio de Janeiro, Brazil, it was a time of great hope. It was there that I witnessed the world's nations come together in an effort to protect the planet's biological richness by signing the Convention on Biological Diversity, the first international agreement to explicitly recognize that conserving wildlife and ecosystems is "a common concern of humankind."

My home country, Canada, stepped forward as the first industrialized nation to ratify the Convention, which commits it and 168 other signatory countries to promote the conservation of biodiversity through strong domestic actions such as establishing new protected areas and enacting new laws to protect and recover endangered species. The signing was a banner moment, one that filled us all with hope.

Though it receives far less attention from the public and policy-makers than other global issues, the precipitous decline and extinction of biodiversity is perhaps the greatest threat to the very sources of the planet's basic life-support systems—clean air, water, soil, and energy. Scientists warn us that we are in the midst of a catastrophic biodiversity crisis on a par with earlier mass-extinction events in the earth's history. Of the species we know, some 17,000 are threatened with extinction, including 12 percent of all known birds, nearly a quarter of all known mammals, and a third of all known amphibians (Vie et al. 2009). Climate change is predicted to sharply increase the risk of species extinction within our own children's lifetime. Indeed, according to the Intergovernmental Panel on Climate Change (2007), around 20 to 30 percent of all plant and animal species assessed are likely to be at increased risk of extinction if the global average temperature rises more than 1.5 to 2.5°C over late-twentieth-century levels.

The immediate threats are well known: destruction of natural habitat such as the old-growth forests needed by the northern spotted owl (*Strix occidentalis caurina*); exotic and invasive species like Sitka mule deer (*Odocoileus hemionus sitkensis*) and other nonnative ungulates introduced into the Haida Gwaii archipelago; over-consumption of wildlife species, including overfishing of salmon (*Oncorhynchus* spp.); pollution of marine ecosystems, including Prince William Sound oil spills; and global climate change that is resulting in the decline of yellow cedar (*Chamaecyparis nootkatensis*) and other trees in the Pacific Northwest (Boyd 2004). But we have only recently begun to fully understand that biodiversity loss at such an unprecedented scale poses major threats to, for example, ecosystem integrity and human welfare (Moola et al. 2007). According to the United Nations' Millennium Ecosystem Assessment (MEA 2005), two-thirds of the direct benefits that people obtain from the "ecosystem goods and services" provided by biodiversity are being degraded or used unsustainably.

The world's remaining temperate and boreal rainforests—such as the Valdivian and Tasmanian rainforests and the Great Bear Rainforest on Canada's Pacific Coast—provide many direct benefits. They are sources of paper and other wood products, as well as medicine, food, and clean drinking water. They also provide habitat for unique assemblages of plants and animals, including a rare white phase of black bear known as the "spirit" bear (*Ursus americanus kermodei*), coastal wolves (*Canis lupus*) that fish for salmon and hunt other marine prey, and canopy-dwelling invertebrates that that live in the tallest of treetop gardens. Such rainforests also provide aesthetic, recreational, and spiritual opportunities and experiences for their human inhabitants and have featured prominently in the culture and traditions of indigenous peoples for millennia. And because they sequester and store billions of tonnes of carbon in their vegetation and soils, they are also a critical shield against global warming. This latter ecosystem service is receiving increasing attention due to growing evidence that protecting carbon-rich ecosystems, especially long-lived rainforests, is an effective strategy to both mitigate and adapt to climate change.

Unfortunately, in spite of their global significance, temperate and boreal rainforests and their ecological importance are poorly appreciated. This has resulted in inadequate levels of protection. For example, nearly all these rainforests have been eliminated from Europe. Just half of the original rainforest cover persists in North America and Chile and only the "guts and feathers" of a once-expansive ecosystem survives elsewhere on the planet. Around the globe, protection levels are far too low (about 13 percent) to sustain temperate and boreal rainforests, particularly in light of a rapidly changing global climate and an ever-expanding human ecological footprint.

The good news is that with strong endangered-species laws, conservation-driven land-use planning, core protection of wildlife habitat in parks and protected areas, and ecosystem-based land-use activities such as FSC-certified logging, we can successfully slow the loss of wildlife habitat and concomitant declines in biodiversity in our temperate and boreal rainforests. Protecting these remaining rainforests is also our best chance of keeping them as carbon sinks. It will promote ecological resiliency and provide broader opportunities for adaptation by species and ecosystems to climate change itself. Indeed, as noted by the Intergovernmental Panel on Climate Change, we must reduce nonclimatic stressors on species and ecosystems if we are going to give them a fighting chance to cope with the effects of global warming.

Temperate and Boreal Rainforests of the World: Ecology and Conservation provides a comprehensive synthesis of the unique ecological and conservation importance of one of the most threatened regions of the planet. The book includes the contributions of leading scientific experts from around the globe to outline an ambitious vision of how we can conserve and manage the planet's remaining rainforests in a truly ecological way that is better for nature, the climate, and ultimately our own welfare.

David Suzuki

LITERATURE CITED

Boyd, D. 2004. Holistic healthcare for biodiversity: from emergency wards to preventative medicine. Pp.1–21 in *Proceedings of the species at risk 2004 pathways to recovery conference*. Ed. T. D. Hooper. March 2–6, 2004, Victoria, BC.

Intergovernmental Panel on Climate Change (IPCC). 2007. Summary for Policymakers. In *Climate change 2007: The physical science basis. Contribution of Working Group I to the Fourth Assessment Report of the Intergovernmental Panel on Climate Change.* Cambridge, UK, and New York: Cambridge Univ. Press.

Millennium Ecosystem Assessment (MEA). 2005. *Ecosystems and human well-being: Synthesis.* Washington, DC: Island Press.

Moola, F. M., D. Page, M. Connolly, and L. Coulter. 2007. Waiting for the ark: The biodiversity crisis in British Columbia, Canada, and the need for a strong endangered species law. *Biodiversity* 8:4–11.

Vié, J.-C., C. Hilton-Taylor, and S. N. Stuart, eds. 2009. *Wildlife in a changing world: An analysis of the 2008 IUCN Red List of Threatened Species.* Gland, Switzerland: IUCN.

Preface

Until now there has been no comprehensive reference from which to build a global understanding of temperate and boreal rainforests and their conservation. In compiling this book, I worked with more than 30 leading scientists to piece together detailed accounts of these remarkable rainforests and elevate their importance among all of the world's unique and rapidly disappearing rainforests. This book includes new map-based technologies and local descriptions of regions poorly understood or completely missed in previous inventories of rainforests, including some surprise additions to the global network of rainforests: the Knysna-Tsitsikamma forests of South Africa, as well as forests in inland British Columbia, western Eurasian Caucasus, Russian Far East and inland Southern Siberia, Eastern Canada, Alps, and northwest Balkans. These forests can now take their place alongside the more recognizable rainforests of Chile, Argentina, Japan, Australasia, and the Pacific Coast of North America as part of a global network of temperate and boreal rainforests.

Perhaps the most compelling reason for doing this book, however, was to shine an international spotlight on the need to conserve the world's rainforests in the face of two looming twenty-first-century threats: rapid changes to the climate that originally shaped these rainforests and humanity's accelerated depletion of the rainforest's finite ecological capital, pushing entire ecosystems to the brink of collapse.

For me personally, the rainforest story begins in Alaska's coastal rainforests, where I first discovered my scientific curiosity and passion for rainforests.

As a young biologist in the early 1990s, I conducted studies on the effects of clearcut logging of old-growth rainforest on neotropical migratory birds, deer, and wolves in the verdant rainforests on Prince of Wales Island, part of Alaska's Tongass National Forest. My summers began each day at the crack of the Alaskan dawn, 3:30 AM for bird surveys, and continued into the late morning. Summers in southeast Alaska are cool and buggy, with very brief interludes of sunshine. Winters were even more challenging, as I searched for deer and wolf prints in the snow while carefully listening for overwintering birds to determine how they survived Alaska's wet, windy winters in a landscape heavily scarred by logging and road-building. During this time, I began to realize that in spite of having a unique firsthand opportunity to witness rainforests in all their splendor, I was helplessly documenting the demise of its coastal giants.

But I was not alone in my concerns. In 1989, Canadian songwriter Bruce Cockburn focused the world's attention on the plight of rainforests—both tropical and temperate—in his inspiring MTV video "If a Tree Falls" (see lyrics below). Cockburn's video was a wake-up call for me personally to join other scientists in a global effort to save rainforests.

Shortly after my Alaskan experience, I began a 13-year career as director of temperate forest programs for the World Wildlife Fund. At the time, I worked with scientists from around the globe to push for a greatly expanded network of protected areas for all the world's forests. In 1997 my efforts were rewarded with a trip to the United Nation's Earth Summit, where I spoke about the need for international agreements to protect the world's rapidly diminishing old-growth or primary forests. Five years later I arrived in the Valdivian temperate rainforest of Chile to take part in international efforts to link rainforest conservation in the "sister" ecoregions of the Klamath-Siskiyou (southwest Oregon and northern California) and Valdivia because both regions were recognized by scientists as among the most diverse temperate forests in the world and faced equally daunting challenges. Despite its uniqueness, the Valdivian temperate rainforests mirror those of Alaska in climate and breath-taking landscape; and like Alaskans, Chileans are struggling to understand the limits of what they can take from rainforests without triggering ecological and social collapse.

Experiences like these prepared me for the first international conference on temperate rainforests in 2003, organized by the World Temperate Rainforest Network.[1] During this conference, I proposed the idea of a global reference work on temperate and boreal rainforests initially envisioned as a collection of scientific accounts and rainforest stories compelling enough to inspire real action by governments and citizens. This book is the culmination of a journey that began in the rainforests of Alaska and continues today through conserva-

---

[1] www.temperaterainforests.org

tion efforts that are spreading across the globe to protect these underappreci-
ated rainforests before they disappear.

Working with other scientists and using the latest inventories and methods,
I attempted to represent the majority of the world's temperate and boreal rainforests. However, while I am confident that the book captures the ecology and
essence of conservation for the vast majority of these rainforests, I hope others
will build on this effort by including areas that we may have missed and further
refine the mapping techniques; and this rainforest story should not be told just
from an ecological perspective. I hope this book will spark a companion piece
detailing the unique aboriginal and responsible forest-management efforts that
are more in harmony with the biological capital of rainforests than industry
practices that have ravished rainforests around the globe; and, of course, I hope
this book will inspire an international call-to-action that puts the conservation
of temperate and boreal rainforests on a par with similar efforts to protect the
world's dwindling tropical rainforests.

I would like to thank a number of individuals for contributing in many
ways to my scientific understandings and conservation values of these rainforests. Pat Rasmussen of the World Temperate Rainforest Network helped to
inspire this book by launching the first international temperate rainforest conference and network (an international consortium of groups working to conserve these rainforests) in 2003. Special thanks to Paul Alaback, who either
contributed to or reviewed many of the manuscripts of this book, including
preparing several of the boxes and figures, and whose early work on temperate
rainforests is reflected in much of thinking throughout this book. Rich Nauman, Jessica Leonard, and Anton Krupicka of the Geos Institute prepared most
of the GIS maps. A grant program from the Environmental Systems Research
Institute provided software support for the GIS maps. David Albert of The Nature Conservancy provided analytical and GIS support for the Tongass conservation assessments. Baden Cross, GIS Analyst at Applied Conservation GIS,
provided data on the inland rainforests of British Columbia. Bruce McLellan,
Wildlife Research Ecologist for the British Columbia Ministry of Forests,
sharpened our thinking through lively debate on caribou management. Anne
Sherrod and Craig Pettitt of Valhalla Wilderness Society shared data and important documents that called our attention to significant omissions in mountain
caribou strategy by the British Columbia government. Wayne McCrory of Mc-
Crory Wildlife Services Ltd. and the Valhalla Wilderness Society provided many
helpful suggestions and edits in Chapter 2. Ecoregional planning conducted by
Maximiliano Bello, Ocean South America, and George Powell, WWF-US,
was integral to the conservation recommendations in the Valdivian temperate

rainforest. Sian Atkinson of the UK Woodland Trust reviewed the European chapter and provided helpful materials. Eric Hosten of Nelson Mandela Metropolitan University tracked down hard-to-find manuscripts about South African forests. Julie Norman, Randi Spivak and Sunya Ince-Johannsen of the Geos Institute helped prepare and proofread tables, boxes, references, and chapter drafts.

Support for this book was provided by Don Weeden of the Weeden Foundation, James Bohnen of the Osprey Foundation, Anna Wiancko of the Wiancko Charitable Foundation, and the Wilburforce Foundation (photos). I am also grateful to several photographers whose work is generously captured in the color photos of this book.

This book is also offered as a standard reference for conservationists and ecologists around the world who have worked for years to study and advocate for the world's temperate and boreal rainforests. While some of these champions are no longer with us, they are remembered for all that they have accomplished: Bill Devall, formerly of Humboldt State University (California), for his devotion to coastal rainforests and his many contributions to the deep ecology philosophy that inspired countless conservationists; Jim Fulton, the first executive director of the David Suzuki Foundation, for his leadership in the campaign to protect the Great Bear Rainforest; Colleen McCrory, founder of the Valhalla Wilderness Society, as the force behind the protection of British Columbia rainforest; and former congressman and career activist Jim Jontz, who in the early 1990s introduced legislation in Congress calling for the protection of the Ancient Forests of the Pacific Northwest that eventually led to the Northwest Forest Plan largely ending the wholesale liquidation of old-growth forests in the Pacific Northwest, USA. Without these contributions, there would be even fewer rainforests to celebrate in this book.

And to my daughter, Ariela Fay DellaSala, and my wife, LeeAnn DellaSala, I hope this book will inspire decision makers to conserve rainforests around the world so that you and others like you will continue to draw sustenance from rainforests and marvel at the *Great Mystery* present within them.

Dominick A. DellaSala

Rain forest
Mist and mystery
Teeming green
Green brain facing labotomy
Climate control centre for the world
Ancient cord of coexistence . . .

Through thinning ozone,
Waves fall on wrinkled earth—
Gravity, light, ancient refuse of stars,
Speak of a drowning—
But this, this is something other.
Busy monster eats dark holes in the spirit world
Where wild things have to go
To disappear
Forever

If a tree falls in the forest does anybody hear?
If a tree falls in the forest does anybody hear?
Anybody hear the forest fall?

    —Bruce Cockburn, *If a Tree Falls*

CHAPTER 1

Just What Are Temperate and Boreal Rainforests?

Dominick A. DellaSala, Paul Alaback, Toby Spribille,
Henrik von Wehrden, and Richard S. Nauman

When most people think of rainforests, they think of lush, tropical "jungles" teeming with poison arrow frogs (*Dendrobates* spp.), toucans (e.g., *Ramphastos sulfuratus*), mountain gorillas (*Gorilla gorilla beringei*), and jaguars (*Panthera* spp.). Tropical rainforests are indeed special places, as they account for over half the terrestrial species on Earth (Meyers et al. 2000) while representing just 12 percent of the world's forest cover (Ritter 2008). Their temperate and boreal counterparts are another story, though, one yet to receive the kind of global recognition rightfully merited by tropical rainforests. Their story is told here, beginning with historical and recent accounts to define and map the temperate and boreal rainforests of the world.

Any discussion of rainforests must begin with what we mean by this term and how we map rainforests. Definitions and mapping standards are the mortar with which scientists visually construct biome delineations such as temperate and boreal rainforests. Consequently, the modeling techniques used in this chapter frame the entire book, as each of the regional chapters is built from the approaches set herein. In cases where it is necessary to deviate from globally based models and maps, explanations are given by regional authors of the book. Nevertheless, we now build on earlier approaches and definitions of temperate and boreal rainforests by providing a standardized modeling approach and a consistent methodology for mapping these rainforests. While it was our original intent that readers of this book would use our approach as the up-to-date standard for defining and delineating temperate and boreal rainforests, we note that this is a work-in-progress requiring further refinement and real-world verification as new data sets become available. Similarly, in Chapter 10, we present standardized mapping techniques aimed at determining just how much of this rainforest biome is in strict protection, a necessary step for developing a unifying vision for rainforests globally and for calling on decision makers to protect these rainforests as we do in Chapter 11. Because the process used in this opening chapter is central to the entire book, we put more emphasis here compared to the regional chapters that follow.

SCIENTIFIC HISTORY OF TEMPERATE AND BOREAL RAINFORESTS

Throughout this book we refer to either temperate or boreal rainforests that differ mainly with respect to latitude, climate, and plant associations. For descriptive purposes we separate these rainforest types in this chapter but refer to them jointly throughout much of the book.

Temperate Rainforests

Temperate rainforests have been recognized in some fashion by ecologists for nearly a century (Köppen 1918; Holderidge et al. 1971; Whittaker 1975; Jarmon and Brown 1983; Veblen 1985; Read and Hill 1985; Omernick 1987; Moore 1990; Hickey 1990; Alaback 1991; Kirk and Franklin 1992; Kellogg 1992, 1995; Gallant 1996; Lawford et al. 1996; Schoonmaker et al. 1997; Moen 1999). Most researchers classify them as distinct biomes based on broad differences in dominant vegetation and/or climate, or as inclusions within larger ecoregions (large areas distinguished by their dominant vegetation, climate, and land form). Yet a simple internet search for "temperate rainforest" yields inconsistencies in mapping locations due to gross differences in definitions and mapping techniques.

An earlier term, "high-latitude rainforest," was proposed by researchers to describe the pan-American portion of the biome (Lawford et al. 1996), since this is the most simple and unambiguous way to define *temperate* as contrasted with *tropical* (low-latitude) rainforests, but "high-latitude rainforests" has increasingly been replaced by "temperate rainforests," which generally have milder climates than boreal rainforests, due primarily to comparatively low latitudes. A number of temperate rainforest subtypes are described later in this chapter in order to distinguish rainforests from one another, and this terminology is used throughout this book.

Boreal Rainforests

The border between boreal and temperate has traditionally been defined as the zone where conifer forests give way to deciduous forests, or, in drier regions, grasslands, roughly equated by Köppen (1918) with the −3°C January isotherm in the south (Tuhkanen 1984). The delineation of boreal versus temperate is blurred in montane regions, where temperate coniferous forest transitions seamlessly to boreal conifer forest. The important thing to note here is that *boreal* is a latitudinal zone and should not be conflated with terms such as *continental*; biogeographers are unanimous in recognizing some high-precipitation oceanic regions as part of the boreal zone. Tuhkanen (1984) compared a wide variety of different approaches to delineating the northern and southern limits of the boreal zone, and in the integrated classification he proposed that several of the rainforest regions treated here as "temperate" would be considered part of the boreal zone. Nonetheless, throughout this book, we use the term *boreal* to describe the cold northern rainforests of what in other studies have been more generally termed *subpolar*. As we will see later, these include the Pacific Coast of North America north of ~55°N latitude (chapter 2), the northern half of the inland rainforest of Northwestern North America (chapter 3), much of the wet forests of Eastern Canada (chapter 4), portions of Norway (chapter 6), and Inland Southern Siberia (chapter 9). Because there is no boreal zone in the Southern Hemisphere, relatively colder areas in this hemisphere are considered subpolar.

In reality, many temperate rainforests straddle the abiotic (nonliving chemical and physical factors) boundaries between temperate and boreal, both latitudinally and altitudinally, and more so for oceanic boreal systems. Thus, these rainforests serve as a phytogeographical bridge, facilitating the exchange of mesic (moist) floral elements among neighboring systems and as corridors of latitude- and slope- related south-to-north, north-to-south and slope-up, slope-down migrations of wildlife during periods of climate change. How much of the forests included in this book is boreal versus temperate depends on which classification system chosen. The fact that highly similar forest-species assemblages can be found on both sides of artificially drawn lines is a topic best reconciled to biogeography debates.

RAINFOREST DEFINITIONS

Where and how to draw the line between temperate and boreal rainforests has changed over time as more and better data have become available regarding these unique rainforests and the conditions that have created them. Several geographers who developed classifications for the world's climate included a category for temperate rainforest based, for instance, on some combination of cool temperatures and high rainfall, or cool temperatures and a small annual range of temperatures (see below). Whittaker (1975) in his classic ecology text *Communities and Ecosystems* also identified a temperate rainforest type. Most of these early efforts separated the Southern Hemisphere forests into a broadleaf evergreen forest type, further complicating a comprehensive global definition. These classifications vary widely in how they portray the distribution of temperate rainforests, and especially what types of temperate rainforests occur on Earth.

The prevailing definition of temperate rainforest began with work in the 1980s, when the environmental group Ecotrust and its collaborators proposed a more precise definition so that more accurate global maps and conservation strategies could be developed (Alaback 1991, 1996; Kellogg 1992, 1995). The first iteration of this work included a definition for these rainforests consisting of: (1) annual precipitation exceeding 1,200 millimeters with 10 percent or more occurring during summer months; (2) mean July temperature of 16°C or less; (3) cool dormant seasons; and (4) infrequent fire that is an unimportant evolutionary factor (Alaback 1991). Soon it became apparent that this definition was too restrictive, and more important, it did not accurately characterize availability of moisture, since there was no direct link between evaporation and the required minimum amount of rainfall. The most biophysically precise method of doing this would be to calculate potential evapotranspiration, which corrects for latitude—with increasing latitude, less precipitation is required to maintain the same humidity levels (Stephenson 1990). Potential evapotranspiration was also later shown to precisely predict the distribution of at least one common rainforest tree in northwestern North America, western hemlock (*Tsuga heterophylla*), even including its distribution in interior rainforests of northwestern North America (Gavin and Hu 2006). In the absence of detailed models and global spatial coverages, a more inclusive definition was proffered by Alaback (1996). In this case, temperate rainforests meeting the original criteria for annual rainfall were divided into four subtypes (or zones, including boreal), analogous to subtypes of tropical forests, based on seasonality of precipitation and annual temperatures:

- *Subpolar*—summer rainfall is above 20 percent of the annual total, summers are cool, and snow is persistent in winter, with mean annual temperature below 4°C.
- *Perhumid*—summer rainfall is above 10 percent of the annual total, summers are cool, and typically transient snow is present in winter, with

mean annual temperature of 7°C. "Cool-temperate" also has been used in this context.

- *Seasonal*—summer droughts and fires can periodically occur, summer rainfall is less than 10 percent of the annual total, with mean annual temperature of 10°C.
- *Warm-temperate*—summer precipitation is less than 5 percent of the annual total, winter snow is rare, drought can occur during any season, and mean annual temperature is 12°C or above (Alaback 1996; Veblen and Alaback 1996; Alaback and Pojar 1997).

The threshold values of temperature and precipitation for each of the forest subtypes was determined by examining climatic conditions in areas along the west coast of North and South America that possessed key ecological characteristics associated with rainforests. This has been the prevailing set of definitional parameters for describing rainforest regions used throughout the chapters of this book.

A NEW GLOBAL RAINFOREST MODEL

Building on concepts from Alaback (1991), we developed a strongly organism/ecosystem–driven model for temperate and boreal rainforests that has identified a very small amount of land surface of the earth within the same biome and sharing climatic characteristics and associated ecological processes that rightfully and generally can be called temperate and boreal rainforest. The processes described herein build on earlier work of rainforest ecologists by providing a broad suite of climatic criteria and a standardized approach to mapping rainforests globally.

In this chapter, we use computer modeling to develop defensible criteria for identifying temperate and boreal rainforests and to locate forests not widely recognized as rainforest but meeting our criteria. Further, we create a computer model with high-resolution climate data and compare it to maps created by regional experts.

Rainforest Distribution Model

This book's chapter authors, from a wide range of rainforest regions, provided locations of sites they considered typical of temperate or boreal rainforest in their area. Based on this input, we used climate data for 117 localities from six regions for the initial modeling step: the Pacific Coast of North America

($n = 55$, mostly coastal); Chile and Argentina ($n = 9$); New Zealand ($n = 10$); Tasmania ($n = 6$); Norway ($n = 15$); and Japan ($n = 22$). These regions were selected because we had localities from collaborators, and because there was little dispute that the locations represent rainforests (especially the Pacific Coast of North America, Chile, and New Zealand). Baseline predictors were extrapolated from a global climate data set (Hijmans et al. 2005); redundancy in the model variables was reduced based on a principal-components analysis of the complete data set. The final model was constructed using a MaxEnt modeling approach (Phillips et al. 2006), consisting only of predictors that improved the model. This yielded 11 discrete climate-related parameters. We used the MaxEnt model since it is known to be more conservative compared to other presence-only models, which tend to overestimate occurrence of a particular variable of interest (in this case, temperate and boreal rainforest).

The model was evaluated with a bootstrapping method (Burnham and Anderson 2002), resulting in strong support of the predictive ability of the model ($AUC = 0.90$; values less than 0.5 indicate no predictive capabilities; see Phillips et al. 2006). Based on 100 repeated runs, we quantified the heterogeneity of the ground-truth climate data set, thus ensuring a demarcation of core zones with a high probability of rainforest occurrence in comparison to areas with a lower probability (for mapping simplicity, only high-probability areas were depicted).

The rainforest distribution model generated four additional regions with climate suitable for temperate and boreal rainforests: the Inland Northwest of North America (figure 1-1, middle-right portion of panel a—inland British Columbia), Eastern Canada (figure 1-1, panel b), Great Britain and Ireland (figure 1-1, western corner of panel d), and portions of the Alps (figure 1-1, lower middle of panel d). Notably, two of these regions have not been widely recognized as rainforest by scientists, including the wettest parts of Eastern Canada, which appeared in some form in all map iterations, and some valleys of the eastern Alps, in particular the Salzburg Alps and mountain ranges of western Slovenia. Interestingly, these regions support rainforest lichen assemblages remarkably similar to those of the Pacific Northwest of North America or coastal Norway.

Two lower-latitude regions often considered rainforest by some (e.g., Kellogg 1992), such as the Colchic (Georgia) and Hyrcanic (Iran) forests of the Western Eurasian Caucasus, and the forests of the southern cape of South Africa, were shown to be in a class of their own compared to the more definitive rainforests of the Pacific Coast of North America and Valdivia. Including these warmer and drier outliers in the model calibration invariably resulted in overestimating the global extent of these rainforests by also including South

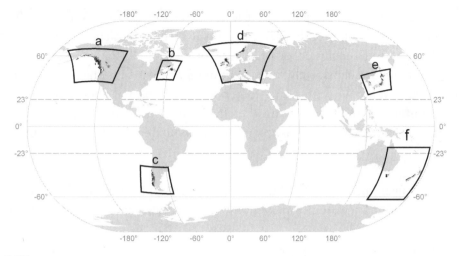

Temperate Rainforest

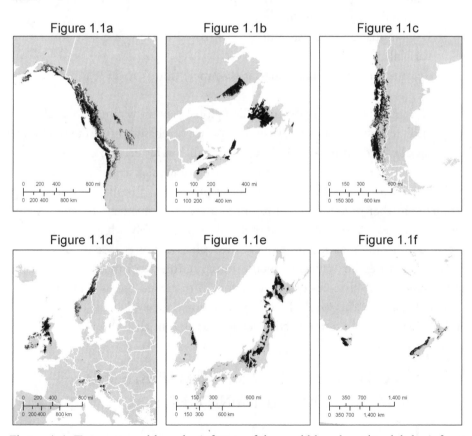

Figure 1-1. Temperate and boreal rainforests of the world based on the global rainforest distribution model, including: (a) Pacific Coast and Inland Northwestern North America; (b) Eastern Canada; (c) Chile and Argentina; (d) Europe; (e) Japan and Korea; and (f) Australasia.

American páramo, high-elevation African equatorial fog forests, and nearly half of the Alps. Retention of the eastern Black Sea region (Colchic), in particular, resulted in model inclusion of large areas of eastern North America, parts of which are indeed climatically similar, but did not agree with our initial criteria on several counts. We settled on a conservative definition of temperate and boreal rainforest based generally on the climate data (see table 1-1; figures 1-2, 1-3) presented for nine regions (some were combined from the set above) as follows:

- Annual (minimum, maximum) temperatures from ~4 to 12°C.
- Annual (minimum, maximum) precipitation from 846 to 5,600 millimeters.
- Snowy winters in high latitudes.
- Significant precipitation (that is, up to 25 percent of annual precipitation) during the driest quarter.
- Low annual temperature fluctuation (based on low annual temperature variability).
- Temperature of warmest quarter (summer) from 7 to 23°C.

This is the first time a spatially explicit global data set was made available for the world's temperate and boreal rainforests that was based on a suite of climate variables obtained from a global data set (available in raster—or grid—GIS format), improvements in computer processing capacity, and statistical models. The model therefore represents an initial cut at producing a global rainforest map, requiring further refinements through the use of regional climate data sets, regional rainforest classifications, and regional maps. Notably, while the minimum precipitation and maximum temperature values reported seem extreme in comparison to earlier rainforest definitions, rainforest communities persist in these regions due to compensatory factors as discussed below and in the regional chapters of this book. This is why regional ground-truth of the model and further study of rainforest classifications are essential.

CLIMATIC PATTERNS OF TEMPERATE AND BOREAL RAINFORESTS

Based on the rainforest distribution model, rainforests were clustered along precipitation and temperature gradients that distinguished them from one another and other forest types.

Table 1-1. Abiotic conditions (climate, altitude) of temperate and boreal rainforests used in the rainforest distribution model, based on a global climate data set.[a]

Model Variable[b]		Norway	Great Britain	Inland Northwest of North America	Eastern Canada	Alps	Pacific Coast of North America	Japan and Korea	Chile and Argentina	Australasia
Mean Diurnal Range (mean of monthly; max temp−min temp; °C)	Min	3.9	5.4	9.2	7.0	6.3	4.3	5.8	4.2	6.1
	Max	7.3	7.6	12.1	11.9	10.4	11.9	11.1	12.1	11.4
	Mean	6.1	6.6	10.8	8.9	8.9	7.6	8.6	7.2	8.9
Isothermality (diurnal temperature/annual range of temperature, °C)	Min	24.1	30.4	27.6	22.3	27.8	24.9	19.2	42.8	44.1
	Max	31.2	42.2	33.2	32.3	34.6	63.6	31.9	60.1	52.2
	Mean	27.9	36.4	30.2	25.9	31.8	33.4	25.1	48.3	47.6
Mean Temperature of Driest Quarter (°C)	Min	-0.9	5.5	-0.1	-11.1	-4.8	-7.0	-14.8	1.2	-1.0
	Max	9.3	13.6	15.2	15.0	15.1	17.9	10.5	16.2	15.2
	Mean	5.4	9.3	3.2	1.7	0.4	9.9	-2.6	9.9	10.0
Mean Temperature of Warmest Quarter (°C)	Min	10.3	10.3	10.8	10.4	10.9	9.9	10.6	6.8	6.8
	Max	14.8	15.2	17.9	18.7	19.5	17.9	22.8	16.3	17.8
	Mean	11.5	12.9	13.5	13.6	15.4	12.5	17.9	11.9	13.2
Precipitation of Driest Quarter (mm)	Min	124	169	106	162	188	6	97	74	143
	Max	447	352	198	399	317	619	487	1378	1169
	Mean	262	248	152	248	245	252	233	499	458

Table 1-1. Continued

Model Variable[b]		Norway	Great Britain	Inland Northwest of North America	Eastern Canada	Alps	Pacific Coast of North America	Japan and Korea	Chile and Argentina	Australasia
Precipitation of Coldest Quarter (mm)	Min	210	244	187	209	209	125	97	230	241
	Max	859	697	458	439	388	1473	506	1403	1803
	Mean	428	431	267	309	269	642	263	801	613
Altitude (m)	Min	25	74	356	86	284	0	69	0	4
	Max	754	842	1487	1319	1548	1263	2612	1066	2334
	Mean	327	314	973	267	826	364	811	278	580

[a]See Hijmans et al. 2005.

[b]Annual mean temperature, temperature annual range, annual precipitation, and precipitation seasonality (coefficient of variation) were included in the model but not in this table as the data are summarized in figures 1-2 and 1-3.

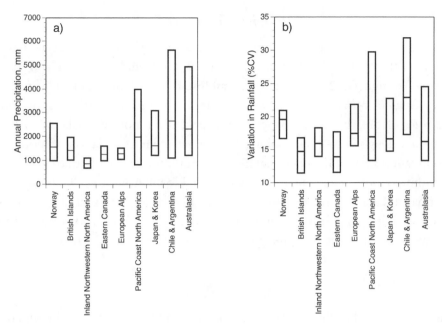

Figure 1-2. Annual precipitation (a) and variation in rainfall (b) of definitive temperate and boreal rainforests, based on a global climate data set (Hijmans et al. 2005) and the rainforest distribution model.

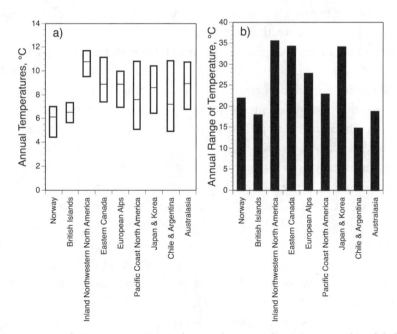

Figure 1-3. Annual temperature (a) and annual range of temperature (b) of definitive temperate and boreal rainforests, based on a global climate data set (Hijmans et al. 2005) and the rainforest distribution model.

Precipitation Gradient

A broad range of annual rainfall amounts occurs in the "classic" temperate rain-forests of the Pacific Coast of North America, Chile and Argentina, and Australasia (see figure 1-2a). As in tropical rainforests, seasonality of precipitation is a key element of rainforest climate that can influence rates of decomposition, the roles of fire, drought, epiphytes, and species composition (Alaback 1996; Losos and Leigh 2004). Just looking at the coefficient of variation of monthly precipitation shows the greatest range in the rainforest regions with the greatest latitudinal ranges (e.g., the Pacific Coast of North America, Chile and Argentina) and also the greatest seasonality, but a less clear pattern in the seasonality of precipitation in smaller regions (see figure 1-2b). More work is needed to clarify how seasonality of precipitation helps effect such differences among rainforest regions.

Temperature Gradient

Based on the global climate data set, Norway had the coolest annual temperature and Inland Northwestern North America (based on southern locales) the warmest (see figure 1-3a). Notably, climate data sets derived from a global reference (Hijmans et al. 2005) may differ from data sets presented in the regional chapters due, for instance, to topographical influences on local climate and the location and density of weather stations.

The comparatively wide range of annual temperatures on the Pacific Coast of North America and in the Valdivian temperate rainforest reflects both its broad latitudinal distribution and a large range in climates from boreal and sub-polar to nearly subtropical. Similarly, the Japanese archipelago spans many climate types (alpine to subtropical, and continental to oceanic), with rainforests distributed zonally.

The annual range of temperature provides a good measure of seasonality of a given region (see figure 1-3b). The regions with the greatest influence from interior climates, such as Inland Northwestern North America, Eastern Canada, Japan, and Korea, all clearly show this influence. The more oceanic climates, such as Norway, the British Islands, and the Southern Hemisphere rainforests, by contrast show a much smaller range of monthly temperatures. This also helps explain why some of the forests in these regions can develop rainforest characteristics with less rainfall than in comparable continental regions.

In sum, rainforests can be grouped both by differences in annual temperature and annual precipitation, with the Inland Northwest of North America the warmest, driest rainforest globally, Norway the coolest (with moderate precipitation), and Chile, Argentina, and Australasia the wettest, with relatively cool-to-moderate temperatures.

OUTLIERS AND OTHER CONSIDERATIONS

The rainforest distribution model did not predict some areas as rainforest which, upon further inspection, showed signs of rainforest conditions or communities. We chose to include some of these as "rainforests at the margins" (or outliers), based on input from regional scientists specializing in the specific regions (see chapter 9). For instance, in some places rainforest communities can persist at precipitation levels lower than the range used in the model as long as there is enough moisture at critical times of the year (e.g., warm summer months) to support moisture-loving species such as lichens and mosses, either directly through some rainfall or indirectly through compensatory mechanisms (e.g., low evapotranspiration rates, high humidity, cool summer nighttime temperatures, and fog). Evidence for this exists for the Knysna-Tsitsikamma forests of South Africa and the Colchic and Hyrcanic forests of the Western Eurasian Caucasus, where persistent fog and high humidity compensate for low summer precipitation and/or hot summers (chapter 9). Such conditions prove suitable for oceanic lichens and humidity-dependent vegetation. The Ussuri taiga of the Russian Far East and the Sayani Mountains of Inland Southern Siberia were too dry for inclusion in the model but have relatively low temperatures and high humidity (chapter 9). Low evaporative losses apparently compensate for drier conditions, allowing humidity-dependent forests to flourish.

The rainforest distribution model also did not identify rainforest in some areas previously suspected to be rainforest. For instance, while Taiwanese montane forests receive sufficient rainfall and cool-enough temperatures zonally (at high elevations) to be considered "temperate rainforest" by some (see Wikipedia[1]; also see Farjon 2005), the lack of a well-defined cool dormant season makes them more ecologically equivalent to cloud or subtropical forests (Alaback 1991), and thus we did not give further consideration to these forests. Iceland's scant boreal forests, though recognized as rainforest by Kellogg (1992), were not included in our rainforest model because the mean annual temperature is below even the minimum used to define rainforests. Icelandic forests also lack the structural complexity associated with temperate and boreal rainforests,

---

[1] www.en.wikipedia.org/wiki/Temperate_rain_forest

such as well-defined canopy layers and gap–phase disturbance dynamics, as trees usually are not long-lived or productive due to severe weather. There are no naturally occurring boreonemoral tree species (see chapter 6) such as elms (*Ulmus* sp.) and oaks (*Quercus* sp.), and there are few of the rainforest lichens common to Norway's rainforests (e.g., *Biatora toensbergii*, *Fuscopannaria ahlneri*, and *Lobaria hallii*).

Although the Appalachian mixed-mesophytic forest of the southeastern United States has been recognized as temperate rainforests by some (see Netencyclo.com[2]; Shanks 1954; see chapter 4), it was not predicted by the rainforest distribution model, presumably because the region has relatively high year-round temperatures and dry summers. However, because there was evidence of rainforest conditions at high elevations (moist pockets of spruce-fir within the larger ecoregion), we briefly mentioned them as a southerly extension of Appalachian boreal rainforests from Eastern Canada that require further study (see chapter 4). In sum, we hope the techniques used here will inspire additional research into these areas in order to further refine our approach.

INTRODUCING TEMPERATE AND BOREAL RAINFORESTS

In the following sections of this book, we discuss seven definitive regions (some regions from above were combined) identified by the model and three outlier regions that collectively make up the global network of temperate and boreal rainforests.[3] We generally organized regions north to south (Western Hemisphere) and west to east (Eastern Hemisphere), as presented sequentially as the book's regional chapters.

Definitive Regions

- Pacific Coast of North America (chapter 2)
- Inland Northwestern North America (chapter 3)
- Eastern Canada (chapter 4)
- Chile and Argentina (chapter 5)
- Europe: Norway, Ireland, Great Britain, portions of the Alps, the Bohemian region, and the Balkans (chapter 6)
- Japan (chapter 7—note that Korea was included in the Russian Far East and Inland Southern Siberia profile, based on author expertise)
- Australasia: Australia, Tasmania, and New Zealand (chapter 8)

---

[2] www.netencyclo.com/en/Temperate_rain_forest
[3] Maps available at www.databasin.org

Outliers (chapter 9)

- Western Eurasia Caucasus (Colchic and Hyrcanic forests)
- Russian Far East and Inland Southern Siberia
- South Africa (Knysna-Tsitsikamma forests)

REGIONAL VS. RAINFOREST DISTRIBUTION MAPS

While the global model was useful in predicting general locations of temperate and boreal rainforests, we often found differences in global projections versus regional delineations made by local experts (see table 1-2). Thus, comparing predicted distributions with regional maps was necessary to ensure that an agreed-upon set of maps was used in the regional chapters. Digital maps for this step were obtained for the Pacific Coast of North America (Kellogg 1995; see figure 4), Inland Northwestern North America (Craighead and Cross 2007), Eastern Canada (described below), Chile and Argentina (provided by Patricio Pliscoff—see below), Australasia (Kirkpatrick and Dickerson 1984), Japan (Miyawaki et al. 1980–89), and Norway (described below). Here, we describe the differences in mapping delineations and reasons for including regional maps, where we had them, in the chapters of the book that follow.

Pacific Coast of North America

Differences in mapping estimates between the global model and regional mapping (Kellogg 1995) were fairly minor (see table 1-2; figure 1-4). The rainforest distribution model yielded a rainforest estimate that was ~9 percent higher than regional mapping (see table 1-2). We present the map by Kellogg (1995) in Chapter 2 because it allowed us to base conservation priorities on regionally specific zones (finer scale) that were not apparent from the coarser rainforest distribution model.

Inland Northwestern North America

The model predicted rainforest to occur on nearly 2.2 million hectares, but only for eastern British Columbia (see figure 1-5; table 1-2). In comparison, using the distribution of western red cedar (*Thuja plicata*) and western hemlock (*Tsuga heterophylla*) (i.e., Interior Cedar Hemlock forests) yielded over 3 times the amount of rainforest at 7.3 million hectares (Craighead and Cross 2007; see chapter 3), with nearly equal amounts in British Columbia and the United States. While the rainforest distribution model and the vegetation-based map showed strong agreement in British Columbia, the Interior Cedar Hemlock

Table 1-2. Global (rainforest distribution model, Kellogg 1992) and regional (based on digital maps from published sources) estimates for temperate and boreal rainforests.

Region	Rainforest distribution model (ha)	(%)	Regionally based estimates[a] (ha)	Kellogg (1992) (ha)	(%)
Pacific Coast of North America[b]	27,274,225	35.0	25,097,930	20,726,700	50.3
Inland Northwestern North America					
British Columbia	2,179,733	2.8	3,879,730		
United States	0	0.0	3,366,874		
Total Inland Northwestern North America	2,179,733	2.8	7,246,604		
Eastern Canada	5,969,641	7.7	6,085,063		
Valdivia					
Chile	12,211,573	15.7	9,752,451	11,675,100	28.4
Argentina	348,371	0.4	2,211,888	323,300	0.79
Total Valdivia	12,559,944	16.1	11,964,339	11,998,400	29.1
European Relicts					
Iceland				195,200	0.47
Norway	4,887,739	6.3	3,747,090	1,459,000	3.5
Great Britain	5,064,759	6.5		1,149,300	2.8
Ireland/Republic of Ireland	1,578,545	2.0		157,300	0.38
Northeast Alps and Swiss Prealps	745,915	1.0			
Bohemia	220,199	0.3			
Southeastern Alps and Northwest Balkans	577,425	0.7			
Total European relicts	13,074,582	16.8		2,960,800	7.2
Japan and Korea	8,295,241	10.6	2,404,404		
Australasia					
Australia	55,989	0.07	1,652,933		
New Zealand	5,458,170	7.0	4,969,590	4,040,400	9.8
Tasmania	3,132,684	4.0	692,300	551,700	1.3
Total Australasia	8,646,843	11.1	7,314,823	4,592,100	11.2
Total Rainforest	78,000,209	1.95[c]		41,177,500	1.1
Outliers[d]					
South Africa (Knysna-Tsitsikamma)	235,483	1.2			

Table 1-2. Continued

Region	Rainforest distribution model (ha)	(%)	Regionally based estimates[a] (ha)	Kellogg (1992) (ha)	(%)
Western Eurasia					
Hyrcanic	1,960,000	10.3			
Colchic[e]	3,000,000	15.8		899,500	2.2
Total Western Eurasia	4,960,000	26.1			
Russia/Siberia					
Russian Far East	6,800,000	35.8			
Inland Southern Siberia	7,000,000	36.9			
Total Russia/Siberia	13,800,000	72.6			
Total Outliers	18,995,483	0.47[c]			
Combined temperate and boreal rainforest total	96,995,692	2.42[c]			

[a]Regional estimates were provided for comparisons to the rainforest distribution model but, due to differences in mapping methodologies, did not include percentages except in the case of Kellogg (1992), which was based on more consistent mapping methodologies.

[b]Differences in rainforest estimates between the two Kellogg references (1992, 1995) are presumed due to refinements in mapping techniques, mainly the addition of the western Cascades in Washington and Oregon, which were not included in the original maps.

[c]Percentages were derived from global forest cover (all forest types) estimated at 4 billion hectares based on FAO (2005) estimates that define forests as >10% tree cover. Plantations are included in estimates.

[d]Outlier estimates, provided by regional authors, were derived from different mapping methodologies not directly comparable to rainforest distribution estimates or other regional estimates.

[e]Kellogg (1992) lists this region as Eastern Black Sea (Turkey, Georgia).

map extends this rainforest type southward for roughly 430 kilometers into northeastern Washington, northern Idaho, and northwestern Montana (see figure 1-5a). Based on local knowledge, we choose the map of Interior Cedar Hemlock forests for Chapter 3.

Eastern Canada

For this region, we overlaid the Thornthwaite (1948) index for perhumid regions (100+ moisture index) onto digital layers of vegetation obtained from coniferous and mixed forest types (source: Canadian Vegetation and Land Cover data set, www.nrcan.gc.ca). This shapefile is based on satellite data obtained in 1995 by the Advance Very High Resolution Radiometer

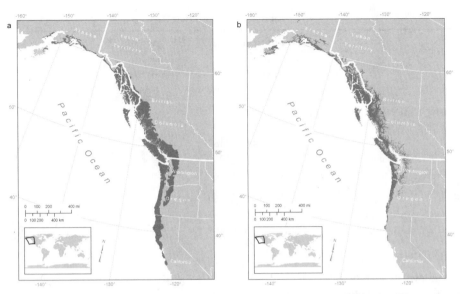

Figure 1-4. Temperate and boreal rainforests of the Pacific Coast of North America based on (a) regional mapping (Ecotrust 1995) and (b) the rainforest distribution model.

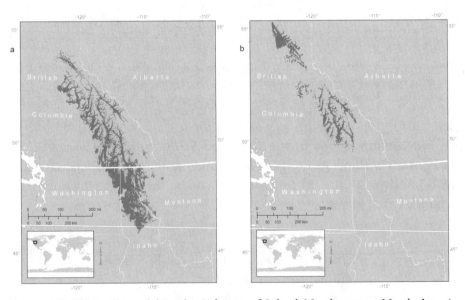

Figure 1-5. Temperate and boreal rainforests of Inland Northwestern North America based on (a) regional mapping (Craighead and Cross 2007) and (b) the rainforest distribution model.

(AVHRR) on board the NOAA-14 (National Oceanic and Atmospheric Administration) satellite. We assumed these forest types were most likely to include important lichen assemblages and rainforest structure that matched perhumid climatic conditions in the region.

Both the rainforest distribution model and regional map (Thornthwaite 1948) yielded nearly identical area estimates (see table 1-2). However, predicted locations of rainforests from the rainforest distribution model vs. regional mapping differed appreciably (see figure 1-6). Thus, we used the regional map in

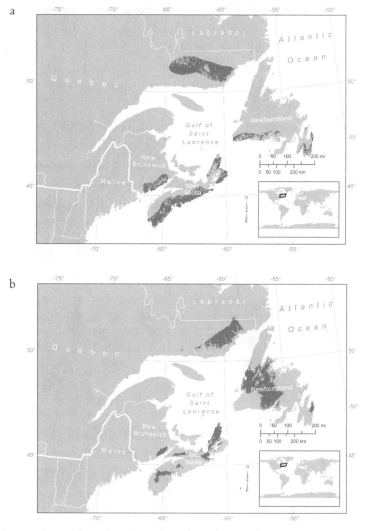

Figure 1-6. Perhumid boreal and hemiboreal rainforests of Eastern Canada based (a) on regional mapping (modified from Thornthwaite 1948) and (b) the rainforest distribution model.

Chapter 4 because it was thought to have higher predictability and greater con-
cordance with forests supporting rainforest lichen assemblages based on local
knowledge.

Chile and Argentina

The primary map source for Chile was the national vegetation survey. This was
originally produced using aerial photography at a scale of 1:50,000 and with
varied level of verification on the ground. Later updates to this information
were produced using Landsat imagery and essentially serve to track loss of forest cover. As a representation of forest cover, the national vegetation survey is
widely used in Chile, is embraced the official source by Chile's Native Forest
Law of 2008, and is fairly reliable. For Argentina no such forest survey exists;
thus we used the same criteria and methods from Chile's national survey and a
series of aerial photos to produce a forest-cover map at 1:500,000 scale without
ground verification.

The rainforest distribution model and regional map yielded similar area estimates for Valdivia (see table 1-2; figure 1-7). However, there were significant
differences in rainforest locations, with the rainforest distribution model extending farther south into the Magellanic (subpolar) rainforests, considered a
separate ecoregion by Chilean scientists (see chapter 5), but missing important

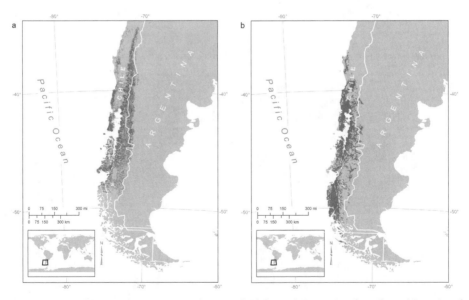

Figure 1-7. Valdivian temperate rainforests of Chile and Argentina based on (a) regional
mapping (digitized from national vegetation surveys) and (b) the rainforest distribution
model.

rainforest locations in the north and in Argentina. Notably, because the Magellanic forests can be considered rainforest by the standards set forth herein, regional authors included some mention of them in Chapter 5. We used the regional map in Chapter 5 because it is widely accepted in regional conservation planning.

Europe

Norway was the only regional map available for comparisons to the rainforest distribution model in Europe. The regional map in this case was based solely on floristic data, namely distribution of epiphytic lichens housed at the Norwegian Lichen Database.[4] Notably, the core area of boreal rainforest in Norway (and Europe) is rather well outlined by the distribution of just two lichens—*Rinodina disjuncta* and *Pyrrhospora (Lecidea) subcinnabarina*—also known from the Pacific Coast of North America (see Tønsberg 1992, 1993; Sheard 1995). The distribution of three other lichens demark the northern and southern limits, with *Lobaria hallii* delimiting boreal forests with occurrences in ravines and by waterfalls, and *Leptogium burgessii* and *Pyrenula occidentalis* the southern boreonemoral (temperate) rainforests.

The rainforest distribution map of Norway estimated about 1.1 million hectares (~30 percent) more rainforest than the estimate generated by regional authors (see table 1-2; figure 1-8). In this case, the rainforest distribution model may have correctly predicted conditions suitable for rainforests but local differences in soils, wind exposure, or human disturbance may preclude rainforest development. Therefore, the Norway regional map was used because it was prepared with regional forest inventories based on known rainforest lichen assemblages (see chapter 6).

Japan

About 5.9 million hectares (over 3 times) more rainforest was estimated by the rainforest distribution model compared to a digitized map of Japan's rainforest zones (see table 1-2); figure 1-9), which were based on finer-scale mapping and therefore used in Chapter 7.

Australasia

About 1.3 million hectares (18 percent, table 1-2) more rainforest was predicted by the rainforest distribution model compared to regional mapping (see figure 1-10). Differences were greatest for Tasmania, where the rainforest

---

[4] www.nhm.uio.no/botanisk/lav/index.html

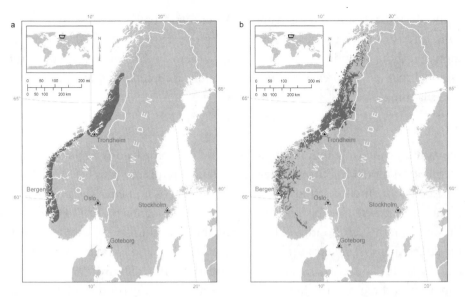

Figure 1-8. Boreal and boreonemoral rainforests of Norway based on (a) regional mapping (derived from lichen distribution maps) and (b) the rainforest distribution model.

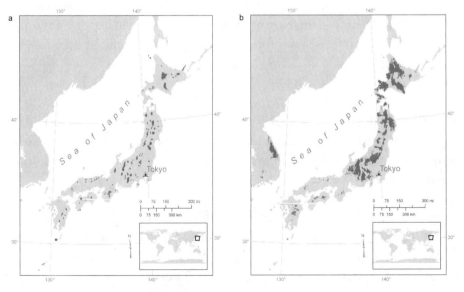

Figure 1-9. Temperate rainforests of Japan based on (a) regional mapping (Miyawaki et al. 1980–1989) and (b) the rainforest distribution model.

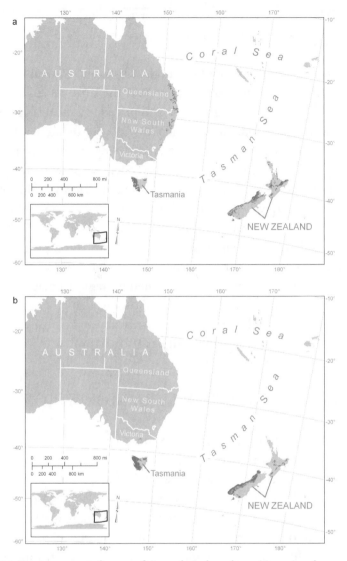

Figure 1-10. Temperate rainforests of Australasia based on (a) regional mapping (Kirkpatrick and Dickerson 1984) and (b) the rainforest distribution model.

distribution model estimated about 2.4 million hectares (over 4 times) more rainforest than the regional map. Conversely, the rainforest distribution map estimated about 1.6 million hectares less rainforest along the Australian coastline (New South Wales). Notably, about 151,173 hectares and 830,769 hectares of the regionally based totals (Kirkpatrick and Dickerson 1984) were classified as clear felled or forests patchily distributed, respectively, at the time. So the overestimate of rainforest by the model may have been partially compensated by the mapping of cleared forests by regional experts. Because the regional maps included more of the Australian coastline where rainforests are known to occur, they were used in Chapter 8.

In sum, the rainforest distribution model was useful in establishing an objective upper range of potential rainforest, was the only standardized data set available for comparisons among regions, and provided a reliable global rainforest total. However, the model had a tendency to overestimate rainforest extent in most, but not all, regions when compared to site-specific mapping and regional expertise. The rainforest distribution model was potentially confounded by human disturbance and local site conditions. Rainforest estimates derived from regional maps, however, also have limitations, as they cannot be compared among regions due to differences in mapping techniques, data sources, and mapping scales. Thus, in making relative comparisons among regions and predicting new localities, the global rainforest distribution model performs quite well; however, for regional specificity we relied on regional maps, as they had a higher degree of reliability at that scale. Follow-up mapping assessments and modeling is recommended in both cases—regionally and globally—to improve rainforest estimates and mapping techniques.

TEMPERATE AND BOREAL RAINFOREST TOTALS

Based on the rainforest distribution model, the Pacific Coast of North America (British Columbia and the United States combined) by far contains the most expansive temperate and boreal rainforests globally, representing over one-third of the world's totals (see table 1-2). Our estimate for this region is notably less than prior estimates (50 percent). Differences are due largely to rainforest areas added in the rainforest distribution model and different mapping techniques, which obviously affected regional totals. Nonetheless, in decreasing order, rainforest extent was then highest for European rainforest relicts (disjunctly distributed); Chile and Argentina; Australasia; Japan; Eastern Canada; and Inland Northwestern North America. However, these percentages do not indicate intactness of rainforests within a given region. For instance, some of the last remaining large blocks of temperate rainforests in the world occur in Valdivia, Tasmania, and New Zealand (see chapters 5 and 8), in comparison to highly fragmented European relicts (see chapter 6); and some of the most intact old-growth rainforests occur in the British Columbia and Alaska (see chapter 2). However, regional totals are not affected by conservation status.

In addition to definitive regions, outliers added nearly 19 million hectares to the global temperate and boreal rainforest total (roughly 0.5 percent), with the Russian Far East and Inland Southern Siberia by far containing the largest (73 percent) expanse and South Africa the smallest (~1 percent, table 1–2).

In sum, our estimate for global temperate and boreal rainforest extent (2.42 percent) was more than twice that of previous estimates (1.1 percent; Kellogg 1992), due largely to additional regions estimated by the rainforest distribution model and differences in mapping techniques. However, some regions (Iceland) previously considered rainforest (Kellogg 1992) were not included here as they do not appear to support rainforest communities. Nonetheless, despite these differences there was considerable overlap in regional estimates, with the net result that temperate and boreal rainforests still represent just a fraction of the global forest cover.

RAINFOREST DIFFERENCES IN THE NORTHERN AND SOUTHERN HEMISPHERES

In this section, we examine major differences in gross rainforest characteristics that can be readily grouped by differences in biogeography between hemispheres where these rainforests are found.

Northern Hemisphere

Temperate and boreal rainforests in the Northern Hemisphere are remarkably similar in species composition, at least at the genus level. The largest of these rainforests in terms of areal extent are dominated by conifers (e.g., Pacific Coastal and Inland Northwest North America, parts of Japan, Norway), usually broadly distributed but closely related species of the pine family, including hemlock, true firs (*Abies* spp.), Douglas-fir (*Pseudotsuga menzeisii*), spruce (*Picea* spp.) or pine (*Pinus spp.*), and species of Cupressaceae, especially red cedars (*Thuja* spp.). Other, smaller regions are dominated especially by beeches (*Fagus* spp.; found in Japan and central European fragments) or beech–spruce mixtures (found, for example, in Norway). In general, temperate and boreal rainforests of the Northern Hemisphere have a dense understory of largely deciduous woody shrubs, a variety of widely distributed (often circumboreal) herbaceous plants and a thick mat of bryophytes (mosses and liverworts), lichens, and many fern species. The broad commonalities among these rainforests make sense from a biogeographical standpoint, since the floras of the Northern

Hemisphere are believed to have been derived in large part from common Tertiary ancestors 60–80 million years ago (see Axelrod 1976).

Southern Hemisphere

In Southern Hemisphere rainforests (southern Chile, Argentina, New Zealand, Tasmania and nearby areas), most trees are broad-leaved evergreens, which form a patchy canopy with many layers beneath the dominant overstory, including a broad diversity of both evergreen and deciduous trees and shrubs. The trees are tall and dense, with small tough leaves (Veblen et al. 1996).

Southern vs. Northern Hemisphere

Southern Hemisphere trees are unlike most of the familiar broad-leaved trees in the North. The "southern beech" or *Nothofagus* trees, for example, are not closely related to beeches of the Northern Hemisphere. They are in their own family (Nothofagaceae) and originated in ancient Gondwana before it split into what have become the small areas of temperate rainforest scattered across the Southern Hemisphere (Veblen et al. 1996). This explains why there are many species of trees that are shared at least at the genus level among rainforests in New Zealand, South America, and Australia (Ezcurra et al. 2008). Another big surprise is in the pine family. While pines, spruces, firs, and related species dominate high-latitude forests of the Northern Hemisphere, this entire family is absent in the Southern Hemisphere (Lusk 2008). The principal tree families shared are the most ancient ones, such as the cedars and cypress species (family Cupressaceae), that were well developed before the continents split apart.

While the Northern Hemisphere is dominated by conifers in the pine family (Pinaceae), trees in temperate rainforests of the Southern Hemisphere belong to a wide assortment of mostly small, specialized families. Among these, the myrtle family (Myrtaceae) is often the most diverse. Some other, more-modern families are also shared between the Northern and Southern Hemispheres, such as the heath and heather family (Ericaceae). In this case, these plants are particularly well adapted to cool, moist conditions, either alpine or subalpine, and have apparently been able to disperse along the Rockies and Sierra Madre in North America down the Andes all the way to Tierra del Fuego. The crowberry (*Empetrum nigra*), for example, has black berries in rainforests of the Northern Hemisphere, but red berries in the Southern Hemisphere (*E. rubrum*), and otherwise looks very similar between hemispheres. The occurrence of these two families may be, in part, attributable to dispersal of the seeds by migratory birds moving between hemispheres, a prospect that also has been proposed for some lichens. A striking exception to the pattern of divergence is the case of an increasing number of possibly relictual lichen lineages being discovered to be shared between the Pacific Coast of North America and Tasmanian and/or Valdivian rainforests (Spribille et al. 2010). However, the overwhelming pattern is one of disparity, with contrasting assemblages recurring with bryophytes, most nonmigratory birds, mammals, fishes, and insects. Why are these forests so taxonomically different between hemispheres? Let's explore some of the leading hypotheses.

Continental Drift and Isolation

This is generally considered the key factor explaining hemispheric differences. While the continents in the Northern Hemisphere were well connected many times in the past, including as recently as a few tens of thousands of years ago during glacial cycles, in the Southern Hemisphere many of the land masses that now have temperate rainforests have been isolated from each other since the late Tertiary period (over 60 million years ago—see Lawford et al. 1996; Veblen et al. 1996; Arroyo et al. 2000). This has lead to adaptive radiation events in species with ancient lineages, resulting in many unique forms (endemics).

Geography

Most of the Southern Hemisphere is dominated by ocean, and at the high latitudes land masses are highly fragmented and have been since the upper Tertiary some 2 million years ago, when the rainforest zone became progressively isolated by xeric climates to the east and north triggered by the uplift of the Andes (Arroyo et al. 1996). Thus, most temperate rainforests have milder winter climates with rainfall evenly distributed over the growing season. This unique climate leads to a more subdued role for wildfire and to a more limited adaptation to extreme cold. Even subalpine species from the Southern Hemisphere are generally not hardy enough to survive in continental rainforests of the Northern Hemisphere (Lawford et al. 1996; Veblen and Kitzberger 2002).

Endemism

The vast majority of species in temperate and subpolar rainforests of the Southern Hemisphere are unique to each continent (South America, Africa, and Australasia), and sometimes to a specific area due to their relictual taxonomic status and long periods of isolation (Lawford et al. 1996; Smith-Ramirez 2004; Hinojosa et al. 2006; also see chapters 5 and 8). By contrast, in the Northern Hemisphere fewer species are limited to specific habitats or areas, although island biogeographical effects in northern coastal latitudes have triggered speciation events at the subspecies level (see chapter 2).

Species Mutualisms

Many species in the Southern Hemisphere evolved from tropical affinities (e.g., Valdivia—see chapter 5), including complex interactions between plants, herbivores, pollinators, and seed-dispersing species. Further, most trees in Southern rainforests produce edible fruits and have co-evolved with seed-dispersing animal species (Armesto et al. 1996). In contrast, most rainforest trees of the Northern Hemisphere are conifers with less direct and specific co-evolution with pollinators and seed dispersers (e.g., Willson et al. 1990).

TEMPERATE AND BOREAL RAINFORESTS VS. TROPICAL MOIST RAINFORESTS

Tropical rainforests, as their name implies, are bracketed by the tropics of Cancer and Capricorn (see figure 1-11; table 1-3). They cover about 6 times more area than temperate and boreal rainforests (~2 percent versus 12 percent of the world's forests). Tropical rainforests are generally drenched in warm, moist climates with little seasonal temperature variation within 1 kilometer of sea level. On the other hand, temperate and boreal rainforests are generally but not

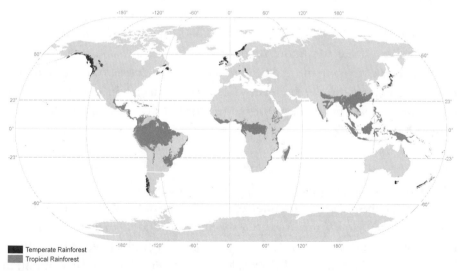

Figure 1-11. Tropical moist (Olson and Dinerstein 1998) and temperate and boreal rainforests of the world.

Table 1-3. General features distinguishing tropical moist rainforests from temperate and boreal rainforests.

Feature	Tropical Moist Rainforest[a]	Temperate and Boreal Rainforest
Distribution	up to 23° latitude from the equator: large belts across South America, Central America, Southeast Asia, and Africa	~30–69° latitude, disjunct, mainly coastal: Pacific Northwest, Alaska, British Columbia, Chile, Argentina, Tasmania, New Zealand, Australia, Japan, Europe
Extent	~12% of present global forest cover, reduced by over half of estimated historic levels	~2% of present global forest cover, reduced by ~half of estimated historic levels
Deforestation (2000–2005)	1–2% annual[b], especially high in South America and Africa, mostly converted to agriculture	forest cover generally increasing, but old growth replaced by tree plantations
Annual Mean Temperature	23–27° Celsius	~4–12° Celsius
Seasonality	uniform temperature with wide variation in rainfall patterns (up to a 3-month dry season)	varied temperatures, snow in winter, greater precipitation in fall and winter with summer rains over 14% of annual precipitation
Moisture	over 1,700 mm, high humidity, high evapotranspiration	846–2658 mm, high humidity, low evapotranspiration
Canopy diversity	multilayered, rich epiphytes (orchids, bromeliads), and abundant lianas	generally multilayered, rich epiphytes (lichens, mosses), lianas less developed
Forest height	20–50 m	10–70 m
Soils	thin litter layer, infertile and severely leached except in volcanic and riparian areas; large nutrient pools in trees	rich humus, highly productive and rich in invertebrates, large amount of coarse, woody debris
Biomass	moderate (100–250 metric tons/ha), highest in dipterocarps (Southeast Asia)	low (Europe) to exceptional (redwoods, Pacific Northwest, Tasmania, Valdivia) (100–1867 metric tons/ha)
Productivity	high-exceptional	exceptional (marine, freshwater, terrestrial)
Nutrient cycling	rapid decomposition rates	slow decomposition rates
Pollination	exceptional	low in conifers
Plant and animal richness	exceptional, over half of terrestrial species on Earth, generally 5–10 times that of temperate forests	low (Europe) to moderate (Japan, Valdivia), but high for mosses and lichens
Endemism	exceptional, many species unique	low (Europe), moderate (California), high (Chile and Argentina)
Tree richness	exceptional (50–200 species/ha)	low to moderate (1–20 species/ha)

[a]Synthesized from Terborg (1992); Richards (1996); Kricher (1997); Myers et al. (2000); and Losos and Leigh (2004).

[b]Deforestation rates based on total forest cover lost on a continental scale (FAO 2005). Individual countries with rainforest, however, may have higher or lower rates of deforestation or show afforestation due to tree planting.

exclusively found along coastlines at middle to upper latitudes, and can extend to nearly timberline (exceptions include Inland Northwest of North America, the Alps, and Inland Southern Siberia).

Climatically, temperate and boreal rainforests have a more distinctive seasonality (especially wider temperature swings), and greater range of precipitation types including snow and sleet, than tropical counterparts (see table 1-3). High temperatures in the tropics lead to high evaporation rates and the development of daily clouds above the forest, so that they can recycle 70 percent or more of their annual rainfall. Temperate rainforests, on the other hand, are cool and wet, with slower rates of decomposition and low evaporation rates. To better understand the differences between these rainforest types, we turn to some key concepts in forest ecology.

Ecologists today generally recognize that forest ecosystems are comprised of three main "ingredients": *composition*—the mix of species in a forest; *structure*—the vertical and horizontal dimensions and spatial patterns of a forest; and *function*—the workings of a forest expressed through nutrient cycling, food-web and disturbance dynamics, forest succession, pollination, and many other processes (Perry et al. 2008). The regions identified as temperate and boreal rainforest in this book have a suite of underlying characteristics along these lines that can be used to further distinguish them from each other as well as from their tropical counterparts.

Structure

Both temperate and tropical rainforests (boreal less so) have complex forest canopies composed of many canopy layers, creating dense and continuous vegetation cover that provides for rich fauna from the ground up. In both forest types, canopy gaps and emergent crowns of dominant trees create complex spatial patterns in the lower strata. A key difference in rainforest canopies is that temperate rainforests are dominated by conifers (except in the Southern Hemisphere, where they are dominated by broadleaf evergreens, and in Japan and Europe, where they can be deciduous), while tropical rainforests are dominated by broad-leaved trees enveloped by numerous lianas (Valdivia, New Zealand, Hyrcanic, and South African temperate rainforests also have lianas). Both rainforest types often have a high degree of standing dead trees (snags) and fallen logs that provide structure and habitat for scores of plant and animal species (Baker et al. 2007; Perry et al. 2008).

Function

Biomass in temperate rainforests is exceptional on a global scale, exceeding that of tropical rainforests (Smithwick et al. 2002; Losos and Leigh 2004; Keith et al.

2009; see table 1-3). For instance, one study of a young temperate rainforest in Oregon showed that it could fix as much carbon per year as some mature tropical rainforests (e.g., 36 metric tons of organic matter per hectare annually—Fujimori 1971). Another study found primary forests in Australia capable of storing up to 1,867 metric tons per hectare, the world's highest known total biomass carbon density (Keith et al. 2009). However, while tropical forests are not exceptionally carbon-dense systems, they still play the dominant role for forest contributions to global carbon cycles due to their high rates of productivity, decomposition, long growing seasons, and the large land area they still occupy.

Evergreen needles (or leaves) are a common characteristic of the vast majority of tree species that grow in temperate and boreal rainforest climates. They allow rainforest plants to photosynthesize throughout the year in most coastal temperate areas, helping to explain the high productivity of these rainforests (Waring and Franklin 1979). The mild climate of these rainforest regions may explain why most of the tallest trees in the world grow there. Examples from around the world include towering *Eucalyptus* forests in southeastern Australia, massive coastal redwoods and alerce in California and Chile, respectively, and ancient coastal Douglas-fir (*Pseudotsuga menziesii* var. *menziesii*) of northern California and the Pacific Northwest. Finally, a continuously mild, wet climate, combined with minimal genetic losses during Pleistocene glaciations, may have played a role in maintaining the rich genetic diversity of conifer species in the Pacific Northwest but led to losses in other regions (Waring and Franklin 1979; Premoli et al. 2000).

Coastal rainforests also are productive places for marine life, with strong linkages between marine and terrestrial ecosystems (Simenstad et al. 1997). Well-known examples include the marbled murrelet (*Brachyramphus marmoratus*) of the Pacific Coast of North America, a coastal seabird that summers at sea but breeds and nests in the tops of old-growth trees; and historical links between Pacific sea-run salmon (*Oncorhynchus* sp.) and terrestrial predators such as bears (*Ursus* spp.) and wolves (*Canis lupus*), which, in the Great Bear Rainforest of British Columbia, prey upon salmon and help fertilize coastal riparian forests through their droppings (see chapter 2).

Composition

Compared to the tropics, in Northern Hemisphere rainforests plant and animal species richness is generally low, and endemism low to moderate, with some noted exceptions (see table 1-3), including island systems (e.g., Cook et al. 2001). However, lichens appear to be much more diversified at high latitudes than in the tropics (witness ~750 species for a single southeast Alaskan rainforest fjord compared to ~550 species in all of Thailand; Spribille et al. 2010). Even if many more lichens are discovered in the tropics and the relative richness gap closes, it appears that the tropics are by no means richer on the orders of magnitude that apply to some other groups of organisms. Outstandingly high levels of species richness also have been documented in basidiomycete fungi ("mushrooms") with hyper-diverse floras documented in coastal rainforests of British Columbia (Roberts et al. 2004) and over 750 macro-fungal species from a single stand of old-growth forest on a hill in rural Victoria on Vancouver Island (Češka 2009). Here, too, numbers may be far higher than in the tropics, especially of ectomycorrhizal fungal species (a type of mycorrhizae composed of a fungus sheath around the outside of root tips—Allen et al. 1995). How these numbers stack up in the long term against species numbers in the more poorly known Tropics remains to be seen, but the fact that key physiological processes for many fungal and lichen species are optimal at cool temperatures through community adaptation (Friedman and Sun 2005) suggests that, for lichens at least, the pattern may hold.

The generally low diversity of trees species in temperate rainforests, with some noted exceptions such as Valdivia and Japan (see table 1-3), should not seem too surprising, since these rainforests tend to have dense overstory canopies and occur in cloudy climates at high latitudes, leaving little light available for understory canopy layers. Many endemic plant species are associated with warm-temperate or seasonal rainforests, such as the forests in south-central Chile and northern California, as well as all rainforests that occur on islands, and other areas in the Southern Hemisphere. In addition, many moisture-adapted taxa that provide a unique physiognomy and structure closely tied to these rainforests, including epiphytic mosses, liverworts, and lichens, are associated with moist rainforest climates (Goward and Spribille 2005; see table 1-3). In these groups, endemism is locally high in Tasmania and New Zealand, Japan, Valdivia, and parts of northwest North America while it is low to nonexistent in the isolated patches of rainforest in Europe and Eastern Canada. This is likely correlated with the extent of glaciation and/or availability of extensive glacial refugia, combined with a long history of good dispersal across and between continents in these regions. Other species-rich taxa in these rainforests include insects (mostly soil and canopy species) and gastropods (mainly in the Pacific Northwest), with high levels of endemism in certain taxa. Apart from that, tropical rainforests are exceptional across taxa (see table 1-3).

Disturbance Dynamics

Stand-replacing disturbances are relatively rare in temperate and boreal rainforests, as they are in tropical moist forests. As a result, both rainforest types are dominated by ancient trees that have a complex structure and pattern, due to a long history of small patch or gap disturbances (see box 1-1). This history, along with the evolution of tree defenses against diseases, has allowed certain tree species to reach very old ages (Waring and Franklin 1979) in not only temperate rainforests (see above examples for tree species) but tropical rainforest trees as well (e.g., *Hymenolobium mesoamericanum* of Costa Rica can live for hundreds of

---

BOX 1-1

Gap Phase Dynamics of Temperate and Boreal Rainforests.

While most temperate and boreal rainforests are subject to various stand-replacing disturbances such as canopy fires, hurricanes, and landslides, forests in moist climates often have small-scale disturbances that serve to maintain the species composition and structure of the forest over time. Some authors have called these disturbances "maintenance dynamics" (see Veblen and Alaback 1996; Perry et al. 2008). A key ecological consequence of frequent gap disturbances is that a wide range of light environments and ecological conditions can be maintained in a forest that enriches its structural and compositional diversity. This also promotes a rich assortment of plant and animal species requiring vastly different light levels (e.g., both shade-tolerant and -intolerant species), and implies forest structure and composition can be theoretically maintained indefinitely. The extent to which a given rainforest is dominated by gap dynamics depends on many factors, including susceptibility to intense windstorms or geomorphic disturbances (landslides and flooding), as well as the susceptibility of individual trees to mortality, insects, and disease.

Key disturbance features of temperate and boreal rainforests are summarized as:

- Usually small-scale events affecting 1–4 percent of the forest area annually, although these gaps are eventually filled by light-seeking plants, creating a continuous push-pull dynamic between gap-dependent and gap-avoiding (anti-gap) species (Nowacki and Kramer 1998; Franklin et al. 2002).
- A small number of trees are killed in each disturbance event, usually fewer than 10 trees (Lertzman et al. 1996; Ott and Juday 2002).
- Gaps vary widely in size and shape, creating a rich mosaic of conditions in the forest (Ott and Juday 2002).

BOX 1-1
Continued

- When gaps are created by wind events, root-throw can create a rich diversity of soils and microhabitat conditions in the forest, including "pit and mound" micro-topography (Bormann et al. 1995) and nesting sites for birds (e.g., winter wrens *Troglodytes troglodytes* often nest in root-wads).
- Tree architecture, including rooting depth, height and exposure of canopy, and resistance to decay fungi play key roles in determining susceptibility to windthrow.
- Openings in canopy created by gaps promote regeneration of tree and understory species, leading to greater diversity in the forest (Spies et al. 1990; Franklin et al. 2002).
- While in theory gap disturbances can maintain the structure and composition of the forest indefinitely, in practice gap dynamics can lead to changes in forests due to changes in the environment at the time of gap creation, including seed availability and dispersal, micro-climate, and specific characteristics of a given gap event (see Lertzman et al. 1996).

---

years—Fichtler et al. 2003). The infrequency of natural fires in both rainforests adds to tree longevity (e.g., see Gavin et al. 2003).

While both tropical and temperate rainforests are affected by and in turn affect regional climates, tropical rainforests, along with the world's oceans, play a major role in the planet's climate regulation. When either rainforest type is cut down, much of their stored carbon is released as carbon dioxide, thus contributing to global warming as well as regional changes in moisture (evaporative losses) and temperature (as discussed in Chapter 11). Understanding this basic fact is key to climate change negotiations for protecting the world's mature forests in both the tropics and temperate zones for their pivotal role in long-term carbon storage (see chapters 10, 11).

RAINFORESTS: GOING, GOING, GONE?

Unfortunately, both temperate-boreal and tropical rainforests have been reduced by at least half their estimated original extent (i.e., before widespread human-related destruction of rainforests—see Bryant et al. 1997; Myers et al. 2000; Ritter 2008; see table 1-3). Logging in the tropics is typically accompanied by the burning of vegetation and conversion of biologically rich forest to agriculture fields also used by livestock. A recent development is the clearing of rainforest to grow crops for biofuels (e.g., Borneo, Malaysia, and forest thinning in the temperate zone). In the tropics, this comes with severe depletion of already nutrient-deficient laterite (acidic) soils due to the leaching of nutrients otherwise held in place by rainforest trees, thus hampering afforestation efforts. Temperate and boreal forests, on the other hand, mainly have been degraded by conversion of biologically rich, older rainforest to simplistic tree plantations, or have been high-graded, where old high-value trees (or forest patches) are removed without providing for adequate rates of regeneration of older age classes or ecological types (as discussed throughout this book).

Notably, some researchers (Kauppi et al. 2006) contend that the world's forests have been increasing over a 15-year period (1990–2005) measured by accruing wood volume, biomass, and captured carbon (growing stock). While this is certainly a positive development, it misses the point about ongoing losses to intact and high-quality forests such as old-growth or primary forests. Globally, very few large, intact primary forests (e.g., "frontier forests") remain (Bryant et al. 1997). In addition, according to estimates provided by the World Wildlife Fund, approximately 13 million hectares of forests are destroyed globally each year mainly in the tropics.[5] But these losses are not just restricted to the tropics. For instance, the United States was recently ranked seventh in the world in deforestation, an annual rate of 215,000 hectares (FAO 2005). These alarming losses come at a time when deforestation (including forest conversion as used here) was second only to fossil-fuel emissions in global contributions to greenhouse-gas pollutants, although growth in emissions from forestry slowed from 1970 to 2004 (IPCC 2007). These forests are not equated by tree farms achieved through planting, as the difference in terms of quality of forest composition, genetics, function, structure, and long-term storage of carbon (and its release by forestry operations) is hard to measure at a global scale, but such comparison is certainly feasible at regional scales through measures of forest quality, remote sensing, and landscape change-detection analysis.

Ongoing consumption of wood products, particularly in the United States, Canada, Japan, and Europe (where per capita consumption levels are highest), will continue this alarming trend of forest conversion in the temperate zone and complete deforestation in the tropics. Recycling, the use of alternative

---

[5] www.worldwildlife.org/climate/northsouthpartner.html

fibers, and improvements in manufacturing technologies are offsetting this trend somewhat. Greater interest in the conversion of cellulosic fiber from forests to liquid fuel (biofuels), however, will put more pressure on the world's forests, both tropical (UNEP 2009) and temperate/boreal (Searchinger et al. 2009). Afforestation cannot keep pace with ongoing demand without further degradation of rainforest biota from the loss of primary forests and the suite of ecosystem services they uniquely provide.

ABOUT THIS BOOK

This book, while focused primarily on the ecology of temperate and boreal rainforests, is intended as a rallying call for global action to conserve these rainforests, which, like so many of the world's rainforests, are at a critical juncture. Each of the regional chapters is a closer examination of the history and ecological characteristics of the largest remaining examples of temperate and boreal rainforest, and provides essential information that can be used to make clearer global priorities for the conservation of these important rainforests.

The regional chapters (chapters 2–9) largely maintain a consistent structure throughout that includes basic information on rainforest location and types, climatic conditions, significant ecological attributes of regional and global importance, ecological processes such as natural disturbances and forest succession, keystone or exemplary rainforest species, regional rainforest classifications (zones or subtypes), threats, and conservation priorities. In Chapter 10, we summarize key findings from each of the rainforest regions in order to stitch together a unifying vision, based on fundamental concepts of conservation biology, for conserving the world's temperate and boreal rainforests. We end the book in Chapter 11 with a call for an international accord to prepare these rainforests for the inevitable consequences of climate change. Most important, we hope that the principles and concepts outlined in this book provide a scientific foundation for expanding rainforest protections around the globe, so that these remarkable rainforests will continue to meet the growing demands of human communities for the life-giving services that these forests have provided to us for millennia.

LITERATURE CITED

Alaback, P. 1991. Comparative ecology of temperate rainforests of the Americas along analogous climatic gradients. *Revista Chilena de Historia Natural* 64:399–412.

————. 1993. North-South comparisons: Managed systems. Pp. 347–48 in *Contrasts in global change responses between North and South America*. Ed. H. A. Mooney, E. Fuentes, and B. I. Kronberg. New York: Academic Press.

————. 1996. Biodiversity patterns in relation to climate in the temperate rainforests of North America. Pp. 105–33 in *High-latitude rain forests and associated ecosystems of the west coast of the Americas: Climate, hydrology, ecology and conservation*. Ed. R. G. Lawford, P. B. Alaback, and E. R. Fuentes. *Ecological Studies* 116. Berlin: Springer-Verlag.

————, and J. Pojar. 1997. Vegetation from ridgetop to seashore. Pp. 69–88 in *The rainforests of home*. Ed. P. K. Schoonmaker, B. von Hagen, and E. C. Wolf. Washington DC: Island Press.

Allen, E. B., Allen, M. F., Helm, D. J., Trappe, J. M., Molina, R., and E. Rincon. 1995. Patterns and regulation of mycorrhizal plant and fungal diversity. *Plant and Soil* 170:47–62.

Armesto, J. J., C. Smith-Ramirez, and C. Sabag. 1996. The importance of plant-bird mutualisms in the temperate rain forest of southern South America. Pp. 248–65 in *High-latitude rain forests and associated ecosystems of the west coast of the Americas: Climate, hydrology, ecology and conservation*. Ed. R. G. Lawford, P. B. Alaback, and E. R. Fuentes. *Ecological Studies* 116. Berlin: Springer-Verlag.

Arroyo, M. T. K., M. Riveros, A. Penaloza, L. Cavieres, and A. M. Faggi. 1996. Phytogeographic relationships and regional richness patterns of the cool-temperate rainforest flora of southern South America. Pp. 134–72 in *High-latitude rain forests of the west coast of the Americas: Climate, hydrology, ecology and conservation*. Ed. R. Lawford, P. Alaback, and E. R. Fuentes. *Ecological Studies* 116. Berlin: Springer-Verlag.

————, C. Marticorena, O. Matthei, and L. Cavieres. 2000. Plant invasions in Chile: Present patterns and future predictions. Pp. 385–421 in *Invasive species in a changing world*. Ed. H. A. Mooney and R. J. Hobbs. Washington, DC: Island Press.

Axelrod, D. I. 1976. *History of coniferous forests*. Berkeley, CA: Univ. of California Press.

Baker, T. R., E. N. Coronado, O. L. Phillips, J. Martin, G. M. F. van der Heijeden, M. Garcia, and J. S. Espejo. 2007. Low stocks of coarse woody debris in a southwest Amazonian forest. *Oecologia* 152:495–504.

Bormann, B. T., H. Spaltenstein, M. H. McClellan, F. C. Ugolini, J. K. Cromack, and S. M. Nay. 1995. Rapid soil development after windthrow disturbance in pristine forests. *Journal of Ecology* 83:747–56.

Bryant, D., D. Nielsen, and L. Tangley. 1997. *Last frontier forests: ecosystems and economies on the edge*. World Resources Institute, Washington, DC. www.wri.org/publication/last-frontier-forests

Burnham, K. P., and D. R. Anderson. 2002. *Model selection and multimodel inference a practical information-theoretic approach*. 2nd edition. New York: Springer.

Češka, O. 2009. A survey of macrofungi on Observatory Hill: Fall 2008 and winter 2008/2009. Unpublished report available at www.geog.ubc.ca/biodiversity/eflora/macrofungi_observatory_hill.html

Cook, J., A. L. Bidlanck, et al. 2001. A phylogenetic perspective on endemism in the Alexander Archipelago of the North Pacific. *Biological Conservation* 97:215–27.

Craighead, L., and B. Cross. 2007. Identifying core habitat and connectivity for focal species in the interior-cedar hemlock forests of North America to complete a conservation area design. Pp. 1–16 in USDA Forest Service Proceedings RMRS-P-49.

Ezcurra, C., N. Baccala, and P. Wardle. 2008. Floristic relationships among vegetation types of New Zealand and the Southern Andes: Similarities and biogeographic implications. *Annals of Botany* 101:1401–12.

FAO. 2005. The global forest resources assessment 2005. www.fao.org/forestry/fra2005/en/

Farjon, A. 2005. *Monograph of Cupressaceae and Sciadopitys.* Royal Botanic Gardens, Kew, UK.

Fichtler, E., D. A. Clark, and M. Worbes. 2003. Age and long-term growth of trees in an old-growth tropical rain forest, based on analyses of tree rings and $_{14}C_1$. *Biotropica* 35 (3):306–17.

Franklin, J. F., T. A. Spies, R. Van Pelt, A. B. Carey, D. A. Thornburg, D. Rae Berg, D. B. Lindenmayer, et al. 2002. Disturbances and structural development of natural forest ecosystems with silvicultural implications, using Douglas-fir forests as an example. *Forest Ecology and Management* 155:399–423.

Friedman, E. I., and H. J. Sun 2005. Communities adjust their temperature optima by shifting producer-to-consumer ratio, shown in lichens as models: I. Hypothesis. *Microbial Ecology* 49:523–27.

Fujimori, T. 1971. Primary production of a young *Tsuga heterophylla* stand and some speculations about biomass of forest communities on the Oregon coast. USDA Forest Service Research Paper PNW-123.

Gallant, A. L. 1996. USGS ecoregions of Alaska map. www.explorenorth.com/library/maps/ecoreg-alaska.html

Gavin, D. G., L. B. Brubaker, and K. P. Lertzman. 2003. Holocene fire history of a coastal temperate rain forest based on soil charcoal radiocarbon dates. *Ecology* 84 (1):186–201.

———, and F. S. Hu. 2006. Spatial variation of climatic and non-climatic controls on species distribution: the range limit of *Tsuga heterophylla. Journal of Biogeography* 33:1384–96.

Goward, T. and T. Spribille. 2005. Lichenological evidence for the recognition of inland rainforests in western North America. *Journal of Biogeography* 32:1209–19.

Hickey, J. E. 1990. Change in rainforest vegetation in Tasmania. *Tasforest* 2:143–49.

Hijmans, R. J., S. E. Cameron, J. L. Parra, P. G. Jones, and A. Jarvis. 2005. Very high resolution interpolated climate surfaces for global land areas. *International Journal of Climatology* 25:1965–78.

Hinojosa, L. F., J. J. Armesto, and C. Villagran. 2006. Are Chilean coastal forests pre-Pleistocene relicts? Evidence from foliar physiognomy, palaeoclimate, and phytogeography. *Journal of Biogeography* 33:331–41.

Holderidge, L. R., W. C. Grenke, W. H. Hatheway, T. Liang, and J. A. Tosi. 1971. *Forest environments in tropical life zones.* Oxford, UK: Pergamaon Press.

Intergovernmental Panel on Climate Change (IPCC). 2007. Synthesis report. Geneva, Switzerland. www.ipcc.ch/contact/contact.htm

Jarmon, S. J., and M. J. Brown. 1983. A definition of cool-temperate rainforest in Tasmania. *Search* 14:81–87.

Kauppi, P. E., J. H. Ausubel, J. Fang, A. S. Mather, R. A. Sedjo, and P. E. Waggoner. 2006. Returning forests analyzed with the forest identity. *Proceedings of the National Academy of Sciences* 103:17574–79.

Keith, H., B. G. Mackey, and D. B. Lindenmayer. 2009. Re-evaluation of forest biomass carbon stocks and lessons from the world's most carbon dense forests. *Proceedings of the National Academy of Sciences* 106 (28):11635–40.

Kellogg, E. L., ed. 1992. *Coastal temperate rain forests: Ecological characteristics, status and distribution worldwide.* Portland, OR: Ecotrust.

———, ed. 1995. *The rainforests of home: An atlas of people and place.* Part 1: Natural forests and native languages of the Coastal temperate rain forest. Portland, OR: An Interrain Publication.

Kirk, R. and J. Franklin. 1992. *The Olympic rainforest.* Seattle: Univ. of Washington Press.

Kirkpatrick, J. B., and K. J. M. Dickinson. 1984. Vegetation of Tasmania 1:500,000. Forestry Commission, Hobart.

Köppen, W. 1918. Klassifikation der Klimate nach Temperatur, Niederschlag, und Jahreslauf. *Petermann's Mitteilungen* 64:193–203.

Kricher, J. 1997. *A neotropical companion.* Princeton, NJ: Princeton Univ. Press.

Lawford, R., P. Alaback, and E. R. Fuentes, eds. 1996. *High-latitude rain forests and associated ecosystems of the west coast of the Americas: Climate, hydrology, ecology and conservation.* Ecological Studies 116. Berlin: Springer-Verlag.

Lertzman, K., G. D. Sutherland, A. Inselberg, and S. C. Saunders. 1996. Canopy gaps and the landscape mosaic in a coastal temperate rain forest. *Ecology* 77:1254–70.

Losos, E. C., and E. G. Leigh Jr. 2004. *Tropical forest diversity and dynamism.* Chicago: Univ. of Chicago Press.

Lusk, C. 2008. Constraints on the evolution and geographical range of Pinus. *New Phytologist* 178:1–3.

Miyawaki, A., ed. 1980. *Vegetation of Japan.* Band 1: Yakushima. Tokyo: Shibundo. (In Japanese with German summary.)

———, ed. 1981. *Vegetation of Japan.* Band 2: Kyushu. Tokyo: Shibundo. (In Japanese with German summary.)

———, ed. 1982. *Vegetation of Japan.* Band 3: Shikoku. Tokyo: Shibundo. (In Japanese with German summary.)

———, ed. 1983. *Vegetation of Japan.* Band 4: Chugoku. Tokyo: Shibundo. (In Japanese with German summary.)

———, ed. 1984. *Vegetation of Japan.* Band 5: Kinki. Tokyo: Shibundo. (In Japanese with German summary.)

———, ed. 1985. *Vegetation of Japan.* Band 6: Chubu. Tokyo: Shibundo. (In Japanese with German summary.)

———, ed. 1986. *Vegetation of Japan.* Band 7: Kanto. Tokyo: Shibundo. (In Japanese with German summary.)

———, ed. 1987. *Vegetation of Japan.* Band 8: Tohoku. Tokyo: Shibundo. (In Japanese with German summary.)

———, ed. 1988. *Vegetation of Japan.* Band 9: Hokkaido. Tokyo: Shibundo. (In Japanese with German summary.)

————, ed. 1989. *Vegetation of Japan.* Band 10: Okinawa/Ogasawara. Tokyo: Shibundo. (In Japanese with German summary.)

Moen, A. 1999. *National atlas of Norway: Vegetation.* Hønefoss: Norwegian Mapping Authority.

Moore, K. 1990. Where is it and how much is left? The state of the temperate rainforest in British Columbia. *Forest Planning Canada* 6:15.

Myers, N., R. A. Mittermeier, C. G. Mittermeier, G. A. B. da Fonseca, and J. Kent. 2000. Biodiversity hotspots for conservation priorities. *Nature* 403:853–58.

Nadkarni, N. M., T. J. Matelson, and W. A. Haber. 1995. Structural characteristics and floristic composition of a neotropical cloud forest, Monteverde, Costa Rica. *Journal of Tropical Ecology* 11:481–95.

Nowacki, G. J., and M. G. Kramer. 1998. The effects of wind disturbance on temperate rain forest structure and dynamics of southeast Alaska. USDA Forest Service, Pacific Northwest Research Station. PNW-GTR-421.

Olson, D. M., and E. Dinerstein. 1998. The global 200: A representation approach to conserving the earth's most valuable ecoregions. *Conservation Biology* 12 (3):502–15.

Omernik, J. M. 1987. Ecoregions of the conterminous United States. Map (scale 1:7,500,000). *Annals of the Association of American Geographers.* 77:118–25.

Ott, R. A., and G. P. Juday. 2002. Canopy gap characteristics and their implications for management in the temperate rainforests of southeast Alaska. *Forest Ecology and Management* 159:271–91.

Perry, D. A., R. Oren, and S. C. Hart. 2008. *Forest ecosystems.* Baltimore, MD: Johns Hopkins Univ. Press.

Phillips, S. J., R. P. Anderson, and R. E. Schapire. 2006. Maximum entropy modeling of species geographic distributions. *Ecological Modelling* 190:231–59.

Premoli, A. C., T. Kitzberger, and T. T. Veblen. 2000. Isozyme variation and recent biogeographical history of the long-lived conifer *Fitzroya cupressoides. Journal of Biogeography* 27:251–60.

Read, J., and R. S. Hill. 1985. Dynamics of Nothofagus-dominated rain forest on mainland Australia and lowland Tasmania. *Vegetatio* 63:67–78.

Richards, P. W. 1996. *The tropical rainforest.* Cambridge, UK: Cambridge Univ. Press.

Ritter, M. 2008. The forest biome. www.uwsp.edu/geo/faculty/ritter/geog101/textbook/climate_systems/tropical_rainforest_1.html

Roberts, C., O. Češka, P. Kroeger, and P. Kendrick. 2004. Macrofungi from six habitats over five years in Clayoquot Sound, Vancouver Island. *Canadian Journal of Botany* 82:1518–38.

Schoonmaker, P. K., B. von Hagen, and E. C. Wolf, eds. 1997. *The rainforests of home: Profile of a North American bioregion.* Washington, DC: Island Press.

Searchinger, T. D., S. P. Hamburg, J. Melillo, W. Chameides, P. Havlik, D. M. Kammen, G. E. Likens, et al. 2009. Fixing a critical climate accounting error. *Science* 326:527–28.

Shanks, R. E. 1954. Climates of the Great Smoky Mountains. *Ecology* 35:354–61.

Sheard, J. W. 1995. Disjunct distributions of some North American, corticolous, vegetatively reproducing *Rinodina* species (Physciaceae, lichenized Ascomycetes). *Herzogia* 11:115–32.

Simenstad, C. A., M. Dethier, C. Levings, and D. Hay. 1997. The terrestrial/marine ecotone. Pp. 149–88 in *The rainforests of home*. Ed. P.K. Schoonmaker, B. von Hagen, and E. C. Wolf. Washington, DC: Island Press.

Smith-Ramirez, C. 2004. The Chilean coastal range: A vanishing center of biodiversity and endemism in South American temperate forests. *Biodiversity and Conservation* 13:373–93.

Smithwick, E. A. H., M. E. Harmon, S. M. Remillard, S. A. Acker, and J. F. Franklin. 2002. Potential upper bounds of carbon stores in forests of the Pacific Northwest. *Ecological Applications* 12:1303–17.

Spies, T. A., J. F. Franklin, and M. Klopsch. 1990. Canopy gaps in Douglas-fir forests of the Cascade Mountains. *Canadian Journal of Forest Research* 20:649–58.

Spribille, T., S. Pérez-Ortega, T. Tønsberg, and D. Schirokauer. 2010. Mining lichen diversity in the Klondike. *The Bryologist*. Forthcoming.

Stephenson, N. L. 1990. Climatic control of vegetation distribution: The role of the water balance. *American Naturalist* 135 (5):649–70.

Terborgh, J. 1992. *Diversity and the tropical rainforest*. New York: Scientific American Library.

Thornthwaite, C. W. 1948. An approach toward a rational classification of climate. *Geographical Review* 38:55–94.

Tuhkanen, S. 1984. A circumboreal system of climatic-phytogeographical regions. *Acta Botanica Fennica* 127:1–50.

Tønsberg, T. 1992. The sorediate and isidiate, corticolous, crustose lichens in Norway. *Sommerfeltia* 14:1–31.

———. 1993. Additions to the lichen flora of North America. *Bryologist* 96:138–41.

United Nations Environment Programme (UNEP). 2009. Toward sustainable production and use of resources: assessing biofuels. UNEP headquarters, Nairobi, Kenya.

Veblen, T. T. 1985. Forest development in tree-fall gaps in the temperate rain forests of Chile. *National Geographic Research*. Spring: 162–83.

———, and P. B. Alaback. 1996. A comparative review of forest dynamics and disturbance in the temperate rainforests in North and South America. Pp. 173–213 in *High-latitude rain forests of the west coast of the Americas: Climate, hydrology, ecology and conservation*. Ed. R. Lawford, P. Alaback, and E. R. Fuentes. *Ecological Studies* 116. Berlin: Springer-Verlag.

———, R. S. Hill, and J. Read. 1996. *Ecology and biogeography of Nothofagus forests*. New Haven: Yale Univ. Press.

———, and T. Kitzberger. 2002. Interhemispheric comparison of fire history: The Colorado Front Range, USA, and the Northern Patagonian Andes, Argentina. *Plant Ecology* 163:187–207.

Waring, R. H., and J. F. Franklin. 1979. Evergreen coniferous forests of the Pacific Northwest. *Science* 204:1380–86.

Whittaker, R. H. 1975. *Communities and ecosystems*. New York: Macmillan and Co.

Willson, M. F., B. L. Rice, and M. Westoby. 1990. Seed dispersal spectra: A comparison of temperate plant communities. *Journal of Vegetation Science* 1:547–62.

CHAPTER 2

Temperate and Boreal Rainforests of the Pacific Coast of North America

Dominick A. DellaSala, Faisal Moola, Paul Alaback, Paul C. Paquet, John W. Schoen, and Reed F. Noss

The world's most expansive stretch of coastal temperate and boreal rainforest is sandwiched between the Pacific Ocean and a chain of coastal mountains spanning 23 degrees of latitude and some 3,600 kilometers. Here, rainforests extend from northern Kodiak Island and Prince William Sound, Alaska (61°N latitude) to just south of San Francisco Bay, California (38°N latitude; see figure 2-1). Along the coastline, rainforest is limited to moist climates extending to as much as 160 kilometers inland in southeast Alaska and adjacent British Columbia to 60 kilometers or less in California and the Pacific Northwest. A secondary belt of rainforest up to 100 kilometers wide occurs along the western slopes of the Cascades from southern British Columbia to central Oregon.

RAINFOREST VITALS AND GLOBAL ACCOLADES

With over a third of the world's temperate and boreal rainforests and some of its most intact watersheds, the rainforests of British Columbia and southeast Alaska are a temperate "Amazonia." Southeast Alaska also boasts nearly a third of the world's old-growth temperate rainforest (Carstensen et al. 2007; see table 2-1); and coastal redwood (*Sequoia sempervirens*) is exceptional in having whole forests that have persisted for over 2,000 years.

Temperate rainforests of the Pacific Coast of North America are among the world's champions of storing carbon, primarily in their massive tree trunks, logs

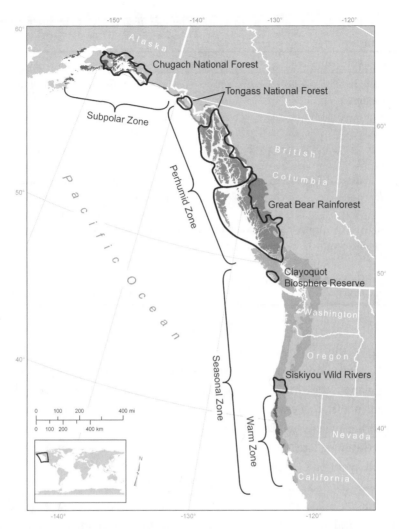

Figure 2-1. Temperate and boreal rainforests of the Pacific Coast of North America, showing rainforest zones (modified from Kellogg 1995) and conservation priority areas. Despite overlap, seasonal and warm temperate (redwood) zones differ in climate and vegetation.

on the forest floor, and thick rich humus layers in the soils (see table 2-1). Coastal redwoods, for instance, store nearly 3,000 metric tons of organic matter per hectare compared with 100–500 metric tons in most tropical and temperate forests, up to approximately 1,000 metric tons in the Pacific Northwest (Fujimori 1971; Beebe 1991; Sawyer et al. 2000; Smithwick et al. 2002), and 1,867 metric tons per hectare in mountain ash (*Eucalpytus regnans*) forests of southeastern Australia and the Central Highlands of Victoria (Keith et al. 2009).

Table 2-1. Globally significant attributes of temperate and boreal rainforests of the Pacific Coast of North America.

Attribute	Importance
More than a third of the world's temperate and boreal rainforests	High conservation value due to global rarity
Some of the largest, oldest trees and most dense carbon pools	Wildlife habitat, long-term carbon storage
High levels of intactness and old growth (northern latitudes)	Habitat for fragmentation sensitive and old-forest dependent species
Expansive mountain fjords, glaciers, and limestone caves	Landscape complexity associated with high levels of beta diversity (turnover in species richness across edaphic gradients)
Exceptional levels of marine, freshwater, and terrestrial productivity, including prolific salmon runs	Rich and abundant wildlife also important for subsistence and cultural values
Exceptional richness of invertebrates, bryophytes, and epiphytes	Food for insectivorous species and nesting platforms for murrelets and other animals

Pacific Coastal rainforests are extraordinarily rich in epiphytes (canopy-dwelling plants), and soil and canopy invertebrates (e.g., see Cooperrider et al. 2000) due, in part, to a cool, wet climate, complex forest canopies, and rich soils (Sillett 1999; Sawyer et al. 2000).

Even the high-latitude rainforests are rich in lichens and bryophytes (e.g., mosses and liverworts—Worley 1972; Schofield 1988; Alaback 1995; Goward et al. 2001; see table 2-1). They also contain 380 vertebrate species, including unique ones (Cook and MacDonald 2001), and support the most prodigious salmon runs on Earth (e.g., *Oncorhynchus* spp. in the Fraser and Skeena Rivers of British Columbia; the Chilkat and Taku Rivers of Tongass National Forest; the Kenai and Copper Rivers of Chugach National Forest; North and Central Coastal British Columbia). Historically, rivers of the Pacific Northwest, especially the Columbia River of Washington and Oregon, and the Klamath River of California and Oregon, had massive salmon runs. Salmon returning to these rivers today represent but a fraction of historic runs, with losses attributed mainly to habitat degradation from logging, hydroelectric dams, agricultural practices, and overfishing in places.

These rainforests also lie at the intersection of land, sea, freshwater, and glacier (ecotones), one of the most productive living tapestries in the world (Ricketts et al. 1999). Rivers transport sediments from glacial melt, known as "glacial flour" for its murky appearance, which aid in the formation of wetlands and de-liver nutrients to aquatic species (Montgomery 1997). Downed rainforest trees embedded in streams offer hiding places and resting pools for migrating salmon in search of cool, shaded waters to spawn. When dislodged, logs float downriver where they are washed ashore, anchoring the beachfront for colonizing vegetation. Depending on stream volume, coastal salinity is diluted at the mouth of rivers, affecting marine productivity (e.g., the massive plume of the Columbia River—Simenstad et al. 1997). Tidal waters flush saline water back through narrow fjords, providing cues for salmon to migrate (Simenstad et al. 1997).

THE MAKING OF RAINFOREST

Pacific Coastal rainforests of North America are dynamic places, shaped by the interplay of geological and glacial forces and ongoing disturbances of various sizes and intensities. Although the retreat of glaciers in the early Holocene some 12,000 years ago opened up vast coastal areas and thousands of nearshore islands for both rainforest species and indigenous people, contemporary rain-forests did not take their present form until some 7,000 to 4,000 years ago (Hebda and Whitlock 1997). Amazingly, these rainforests have only been around for just a few generations of their oldest trees (e.g., coastal redwoods), and in Alaska some rainforest communities (Prince William Sound, Yakutat and Kodiak Islands) have been present for less than 1,000 years.

Pacific Coastal rainforests are surrounded by expansive mountain fjords and glaciers from British Columbia northward. Tidal glaciers are especially pro-nounced in high-latitude areas (LeConte is the southernmost tidewater glacier at 56°N latitude) where for thousands of years they carved U-shaped valleys (mountain fjords) and fingerlike bays, projecting deep into rainforest interior and once extending 80 kilometers or more off the contemporary outer coast-line. Despite this massive accumulation of ice, several regions escaped glaciation, including Haida Gwaii (Queen Charlotte Islands), Brooks Peninsula on northern Vancouver Island, the outer coast of Glacier Bay, Dall Island (southeast Alaska), Kodiak Island, and forests south of the Olympic Peninsula. These areas provided refugia for rainforest species during the Pleistocene glaciations, in-cluding endemic subspecies found nowhere else on Earth (Alaback and Pojar 1997; Cook and MacDonald 2001; Coast Information Team Report 2004). Pacific salmon also include many genetically unique stocks (Allendorf and Waples 1996).

RAINFOREST GRADIENTS AND CLIMATE

Because this coastal rainforest spans such great north-south distances, differences in rainforest communities are especially pronounced. Ecologists like to group these differences along environmental gradients determined by abiotic (chemical and physical) factors underlining the distribution of living communities (biotic factors).

Distributional Gradients

In the most northerly rainforests, tree line dips down below 200 meters elevation and conifers include just two species: Sitka spruce (*Picea sitchensis*) on Kodiak Island and mountain hemlock (*Tsuga mertensiana*) on the Kenai Peninsula. Conversely, conifer richness and endemism in southern locales are much higher, particularly where coastal and inland areas overlap (Alaback 1995; DellaSala et al. 1999). In these southern areas, tree line is sometimes present at the tallest peaks (up ~2,744 meters elevation in places) where alpine tundra takes over. Terrestrial productivity also increases southward, owing to long growing seasons and high nutrient availability. Despite the high productivity of southern forests, forest landscapes become extremely patchy and water-limited, with the southernmost forests restricted to coves and riparian draws until Mediterranean scrub vegetation eventually dominates.

While species richness is generally low in northern latitudes, some species such as bald eagle (*Haliaeetus leucocephalus*) and brown bear (*Ursus arctos*) attain exceptional densities due, in part, to a superabundant food source—salmon— and relatively high levels of intact forest in places. Species that are well adapted to cool, moist conditions, such as lichens, liverworts and mosses, as well as many marine invertebrates, typically have greater diversity in these northern ecosystems than they do in southern areas (Schofield 1988; Alaback 1995; Simenstad et al. 1997).

Rainforest Climate

The collision of saturated marine air against coastal mountains soaks these rainforests in life-giving precipitation, with some locales receiving up to 5,300 millimeters annually (Alaback 1995; Western Regional Climate Center 2006). The regional climate lacks extended warm or cold periods, has a high frequency of cloud cover (60–80 percent monthly), and fog (especially in the redwood region; Redmond and Taylor 1997). In the most northerly locales, rainfall is persistent year round (over 25 percent annual precipitation in summer). Southward, summers become progressively drier, especially in the redwood region, where under 2 percent of annual rainfall occurs in summer months, rendering these forests as more seasonal rainforests (Alaback 1995). Climate is increasingly continental inland, as the moderating influence of the Pacific Ocean diminishes, and snow, uncommon throughout, occurs at higher elevations and latitudes. Precipitation is highest on the windward side of mountains and oceanic islands, with rain shadows on the leeward side.

This persistent moisture has several direct implications to rainforest ecology. The most significant factor is the reduced role of fire, in contrast with drier forest types. South of the Canadian border in the seasonal rainforests of the Pacific Northwest, infrequent fire, every 300–600 years, plays a key role in establishing long-lived species such as Douglas-fir (*Pseudotsuga menziesii*) (Veblen and Alaback 1995; Franklin et al. 2002; Gavin et al. 2003). Farther north in the perhumid and subpolar rainforests (see classifications below), fire has only a localized influence. The other key implication of a persistently humid climate is the development of thick organic layers on the forest floor and slow rates of decomposition. This ultimately leads to the development of peat bogs in northern forests.

ISLAND BIOGEOGRAPHY

The northern rainforest region contains thousands of small islands and some of the largest (e.g., Prince of Wales Island, southeast Alaska, is the fourth largest in the United States) and most mountainous ones (elevations above 1,800 meters) in North America. This is a place where island biogeography, the size of individual islands and their distance from the mainland and one another, combine to influence the composition of rainforest communities. In southeast Alaska, brown bears are distributed along the mainland coast and northern islands of the Alexander Archipelago (the so-called ABC islands—Admiralty, Baranof, Chichagof, and smaller adjacent islands) but are absent on the southern islands (south of Fredrick Sound) that are inhabited instead by black bears (*U. americanus*) and wolves (*Canis lupus*), which also occur on the mainland coast. The Alexander Archipelago of southeast Alaska and the Queen Charlotte Islands contain endemic subspecies of invertebrates, birds, and small mammals, some of which maybe especially vulnerable to extinction due to insularity effects (Cook and MacDonald 2001; Smith 2004).

Other unique rainforest inhabitants include: the Alexander Archipelago wolf (*C. l. ligoni*) from Yakutat Bay to Dixon Entrance of southeast Alaska

except for the ABC Islands; ermine (*Mustela erminea haidarum*), a threatened for-
est dwelling, weasel-like mammal restricted to the Queen Charlottes of British
Columbia; and spruce grouse (*Falcipennis canadensis isleibi*) on Prince of Wales
Island, Alaska. Notably, coastal wolves of British Columbia are highly distinct
and representative of a unique ecosystem, whereas gray wolves of interior
British Columbia are more similar to adjacent populations of wolves located in
Alaska, Alberta, and Northwest Territories (Muñoz-Fuentes et al. 2009). Given
their unique ecological, morphological, behavioral, and genetic characteristics,
gray wolves of coastal British Columbia should be considered an Evolutionary
Significant Unit (ESU) and, consequently, warrant special conservation status
(see Muñoz-Fuentes et al. 2009). Many plant and tree species also are isolated
on islands in British Columbia and Alaska, including alpine and whole forests of
subalpine fir (*Abies bifolia*) and silver fir (*Abies amabilis*) (Alaback 1995; Fedje
and Mathewes 2005).

Another key element that leads to heterogeneity of plant communities and
consequently high levels of biological diversity is the patchy and complex nature of the geological formations, particularly across the archipelago portion of
the region. Most notable is the existence of belts of karst, or limestone deposits,
which occur on several southern islands in Alaska (especially Prince of Wales Island) and Princess Royal Island in British Columbia, where they are associated
with extensive cave systems, many of which contain fossilized ancient carnivores and the remains of other extinct animals. Karst deposits provide sites for
many unique mosses and vascular plants as well as for what were the most historically significant patches of large Sitka spruce in southeastern Alaska (Alaback 1995; Carstensen et al. 2007).

Because much of the northern portion of this rainforest region is insular, island biogeography, combined with human-related disturbances, may increase
extirpation risks of rainforest subspecies (see Hanley et al. 2004; Cook et al.
2006). Other vulnerable species include the Alexander Archipelago wolf, the
Prince of Wales flying squirrel (*Glaucomys sabrinus griseifrons*), ermine (*Mustela
ermine*—several endemic subspecies), and American marten (*Martes americana
caurina* and *M. a. americana*) (Hanley et al. 2004; Flynn et al. 2004, Cook et al.
2006). Additional rainforest biota may be sensitive to human impacts due to
their limited dispersal (e.g., flightless arthropods, endemic mollusks, mosses, and
lichens) and dependence on older forest structures (e.g., downed decayed logs,
tree-fall gaps). In fact, recent research has shown that even remnant patches of
rainforest are critical to the survival of species sensitive to forest fragmentation
(Moola et al. 2004; Halpern et al. 2005).

FOREST SUCCESSION AND DISTURBANCE DYNAMICS

Coastal rainforests of North America are shaped by natural disturbances ranging from small gaps in the overstory canopy created by the death of individual
trees to stand- and landscape-level insect outbreaks, high-wind events (especially in archipelagos—Nowacki and Kramer 1998), and landslides, although
infrequent but severe fire is a disturbance agent mainly south. Landslides deliver
massive amounts of sediments to streams, while extensive floods inundate rainforest areas, creating nutrient and energetic exchanges that influence species
composition, food-web dynamics, and ecosystem productivity (Veblen and Alaback 1995; Alaback and Pojar 1997).

In the moist and cool rainforests of the north, fire is nearly absent and rainforests persist relatively unaltered for centuries. The transitional and seasonal
rainforests to the south; however, have less old growth, owing to infrequent but
large fires and higher logging levels. Old-growth rainforests in the Oregon
Coast Range, for instance, historically accounted for ~25 to 75 percent of the
forest age classes, depending on disturbance dynamics (Wimberly et al. 2000).
Farther north, nearly all (roughly 90 percent) of the productive rainforests in
southeast Alaska and the north coast of British Columbia were old growth, affected little, if at all, by fire (Gavin et al. 2003; Albert and Schoen 2007a). Small-
scale windthrow is the dominant natural disturbance agent in these forests
(Lertzman et al 1996). Large-scale blowdown and landslides on steep slopes and
wind-exposed terrains also occur (Veblen and Alaback 1995; Alaback and Pojar
1997).

Following stand-replacing disturbance events, plant communities proceed
along a successional continuum propagated initially by fast-growing, colonizing
forbs, grasses, shrubs, and conifer seedlings. This early seral stage is then followed
by young, densely packed conifers ("stem-exclusion phase") that overtop and
shade out most understory plants, usually within 15–20 years, and by several intermediate stages leading to old-growth forest (Alaback 1982, 1984; Alaback
and Juday 1989; Franklin et al. 2002; Spies 2004). Barring large disturbances, the
progression to old-growth forest takes some 150–400 years, depending on site
conditions (Alaback 1982; Franklin et al. 2002; Spies 2004; VanPelt 2008). Due
to the recurrence of small disturbances, old-growth rainforests are patchy
places, with pioneering stages persisting at finer spatial scales (i.e., fine-grain
heterogeneity) wherever disturbance resets nature's successional clock. For instance, blowdown of a giant conifer creates a gap in the tallest tree crowns for
life-giving sunlight to penetrate deep within the forest understory, allowing

fast-growing plants to fill the gap and creating a structurally diverse forest from the ground up (Franklin et al. 2002).Therefore, an individual old-growth forest site (or stand) represents a perpetual tug-of-war between gaps created by the death of overstory trees and light-seeking plants racing to fill them (also see discussion of maintenance disturbances in chapter 1). Such fine-scale processes make older forests much more structurally complex and biologically rich than the vast areas of biologically impoverished tree plantations replacing them (Franklin et al. 2002; Lindenmayer and Franklin 2002; Spies 2004).

In general, older forests and large-tree stands tend to have very high levels of species richness and exceptional structural complexity (Lindenmayer and Franklin 2002; Franklin et al. 2002; Spies 2004).They also store vast amounts of carbon important in climate regulation (Smithwick et al. 2002; Luyssaert et al. 2008; Keith et al. 2009).While most conservation efforts rightfully have focused on protecting older forests (Schoen et al. 1988; Strittholt et al. 2006), unlogged, early-successional rainforest communities created by natural disturbance events also are exceptionally rich (Franklin and Agee 2003; Swanson et al 2010).The tail ends of the rainforest successional continuum (natural young and old) include high levels of species richness, complex ecosystem functions, and important forest structures (especially biological legacies) that are either absent or present in much lower levels in tree plantations (see table 2-2). Because they are managed intensely for fiber production, tree farms often include chemical inputs (herbicides, fertilizers, pesticides), and their ongoing maintenance represents chronic ecosystem stressors (e.g., runoff of sediments facilitated by roads and logging on steep slopes that pollute streams).

Other examples of rainforest communities that are of conservation interest yet are not old-growth are periglacial forests, which have developed in the wake of melting glaciers since the Little Ice Age, and shoreline forests that have developed on uplifted beaches in northern southeast Alaska and Prince William Sound (e.g., Carstensen et al. 2007). These are the first forests that have developed since glaciers or the ocean scoured soils from bedrock or sediments. Their soils are poorly developed, and they are quite sensitive to logging and other disturbances that could reduce nutrient availability.

KEYSTONE SPECIES

Keystone species drive the abundance, distribution, and ecological requirements of other species within the rainforest community. Perhaps no species typifies this role better than Pacific salmon (Willson and Halupka 1995; Gende

Table 2-2. Comparisons of old-growth rainforest, naturally regenerating rainforest, and tree plantations along the Pacific Coast of North America.

Type	Feature	Importance
Old-growth rainforest	continuous, multilayered canopy	habitat for wildlife such as neotropical migratory birds, spotted owls, shading for fish
	abundant, large, old trees	sites for owls, murrelets, and lichens; carbon storage
	abundant dead trees (snags)	nesting and foraging sites, anchor soils
	abundant logs	fish habitat, soil stabilization, "nurse" logs for conifer seedlings, salamander and invertebrate sites, soil richness, mycorrhizae fungi
	diverse shrubs/forbs	habitat for ground-nesting birds and mammals, especially deer
	small canopy gaps	structural complexity, habitat for early seral species
Young, naturally regenerating rainforest[a]	biological legacies (logs/snags/large trees) carried over from old forest following distrubance	habitat for old-growth associates, "anchor" soils, shade conifer seedlings from intense sunlight, sites for mycorrhizae fungi
	exceptional shrub and forb richness	habitat for deer and other early seral species, nutrient cycling
	complex vertical and horizontal structure	habitat for rich array of early-seral and some late-seral species
	complex nutrient cycling and energy (food-web) pathways	productive soils and high species diversity
Tree plantations[b]	young, densely packed trees	simplistic vertical and horizontal structure due to dense spacing of small trees
	low understory light levels	sparse forb and shrub layer
	simplistic genomes with seedlings grown to match local site conditions	reduced resilience to climate change
	low levels of nutrient cycling and mycorrhizae fungi	diminished below-ground processes
	lack of biological legacies	impoverished wildlife habitat and ecological processes
	chronic erosion and depletion of soils, particularly from roads and use of heavy machinery	long-term diminished site productivity and/or use of fertilizers and other chemicals to supplement nutrient deficiencies
	high risk of severe fires in drier areas due to high fuel loads and roads (ignition factors)	altered successional pathways

[a]Generated by natural disturbance events such as blowdown, volcanic eruptions, wildfires, and landslides.

[b]Degree of differences vary depending on scale and intensity of forest management.

et al. 2002). More than 190 species of plants and animals eat salmon, including marine mammals, birds, bears, and, surprisingly, even wolves (Cederholm et al. 2000). By way of their dual oceanic and riverine life cycles and their importance to predators, salmon uniquely span marine, freshwater, and terrestrial systems. During salmon runs, for instance, coastal bears derive up to 90 percent of their total annual dietary requirements as well as essential fat stores for hibernation (Temple 2005). Wolves are seasonally dependent on salmon as a primary source of food in this region (Darimont et al. 2008). Amazingly, up to 80 percent of yearly nitrogen uptake in old trees is from nutrients derived from rotting, spawned-out salmon carcasses (Reimchen 2000).

From a conservation standpoint, the potential extirpation of keystone species like salmon could have cascading ecological effects that reverberate through rainforest communities (Lichatowich 1999). For instance, salmon declines are known to reduce nutrient levels affecting food-web dynamics in streams (Gresh et al. 2000) and have the potential to adversely affect regional economies that depend on productive fish runs (Sisk 2007a). Notably, nearly one of four salmon populations is at risk of extinction in this region (Augerot 2004), and several have vanished already (Nehlsen and Lichatowich 1997; Price et al. 2008).

REGIONAL RAINFOREST CLASSIFICATIONS

To account for variability in species assemblages and broad climatic differences across such an expansive region, ecologists have subdivided coastal rainforests using various classifications. For instance, Gallant (1996) developed an ecoregional classification system based on land use, land surface form, potential natural vegetation, and soils, delineating two Alaskan ecoregions that support coastal rainforests: Pacific Coastal Mountains and Coastal Western Hemlock–Sitka Spruce Forests. Nowacki et al. (2001) refined this approach (finer scale) by mapping five areas with coastal rainforest: Alexander Archipelago, Boundary Ranges, Chugach-St. Elias Mountains, Gulf of Alaska Coast, and Kodiak Island. Ricketts et al. (1999) developed ecoregional classifications based on broad differences in vegetation, climate, and landform, identifying five ecoregions in North America with coastal temperate rainforests: North Pacific Coastal Forests (southeast Alaska, Prince William Sound, and eastern Kodiak Island), British Columbia Mainland Coastal Forests, Queen Charlotte Islands, Central Pacific Coastal Forests (Vancouver Island south to southern Oregon), and Northern California Coastal Forests.

In this chapter, we used the classification system of Alaback (1995) and Alaback and Pojar (1997) described in further detail in Chapter 1 primarily because these classifications were developed specifically for this region, were based largely on differences in summer rainfall and temperature, and were useful in grouping conservation priorities. The four zones (see figure 2-1) include:

- *Subpolar*—highest-latitude coastal forests
- *Perhumid*—northern Vancouver Island, British Columbia to southeast Alaska
- *Seasonal*—southern Oregon to central Vancouver Island
- *Warm*—San Francisco Bay to southern Oregon coast.

Perhaps the most conspicuous occupants spanning all rainforest zones, however, are the conifers. Species composition varies but the predominant conifers are Sitka spruce, western hemlock (*T. heterophylla*), western red cedar (*Thuja plicata*), Douglas-fir, silver or amabilis fir (*Abies amabilis*), shore pine (*Pinus contorta*), yellow cypress or yellow cedar (*Chamaecyparis nootkatensis*), and coastal redwood (south).

CONSERVATION PRIORITIES

We nested conservation priorities and rainforest threats within each of the four rainforest zones. Threats varied by zones but, in general, logging was greatest southward (but moving north) and industrial tourism, mining, and salmon farming more prevalent northward (see box 2-1).

SUBPOLAR (BOREAL) RAINFOREST CONSERVATION PRIORITIES

The subpolar zone includes high-latitude rainforest with distinct subpolar attributes (Alaback and Juday 1989; Alaback 1995). Forests are bracketed by glaciers interspersed among alpine meadows, muskegs, and other wetlands and are dominated by mountain hemlock and Sitka spruce. Due to the short growing season and poor soil conditions, conifers rarely exceed 30 meters tall and are found on protected slopes where soil drainage is good (Alaback and Pojar 1997). Annual precipitation is often greater than 3,810 millimeters and snow is more common than in southerly latitudes.

More than 40 percent of rainforests throughout this region have been fragmented by land-use activities, with highest fragmentation levels southward (see Cascadia Scorecard 2007; sightline.org[1]); logging has been advancing northward more recently. The amount of protected rainforest varies from approximately 5 percent that is strictly protected in the Pacific Northwest to nearly 85 percent of the Chugach National Forest in Alaska, but generally below representation targets for most areas. Types of protection range from Wilderness (strictly protected) and National Parks (although commercial recreation and even some development are allowed in Canadian parks) to conservation reserves (some or no commercial logging allowed but other development activities, such as mining or road building, are permissible). Mining and energy development (e.g., Prince William Sound) are ongoing risks to fish and wildlife. Industrial tourism, especially along the inside passage of Alaska and Prince William Sound is increasing. (Tourism is also a benefit to local economies when properly managed.) Fish hatcheries and salmon farming compete with and adversely affect wild fish runs, and overfishing in nearshore waters is reducing fish runs in portions of British Columbia. Extirpation of grizzly bears and wolves (e.g., south of Canada) has altered food-web dynamics through removal of apex predators. Although trophy hunting of wolves, bears, and other large carnivores is legally allowed throughout most areas north of the Canadian border (including most parks and protected areas), such mortality can be managed, whereas the poaching and killing of large carnivores by people for defense of life or property is relatively unpredictable.

Rainforest conservation is hampered by weak wildlife regulations and laws (e.g., British Columbia Forest Range and Practices Act and limitations on the Canadian Species At Risk Act), as well as inconsistent policies (e.g., uncertain federal roadless-area policies in the United States). Other threats include exotic-species invasions (particularly south), rapid growth and expanded access for off-highway vehicles in previously inaccessible areas, and, perhaps greatest of all, climate change. Cumulative impacts have led to loss of salmon runs from southern British Columbia southward, and the imminent extirpation of the northern spotted owl in southwest

---

BOX 2-1
Continued

British Columbia and the Olympic Peninsula of Washington. Global warming is melting tidal glaciers and shrinking mountain snowpack (Mote et al. 2005), and is expected to drop river flows and produce water temperatures too warm for spring and summer salmon runs. Other antic-ipated climate-change impacts include increased erosion from floods and landslides.

---

[1]www.sightline.org/maps/maps/forests_over_cs04m

The subpolar zone reaches its western terminus on Afognak Island, in the northern portion of the Kodiak Archipelago, and its northern terminus in the Prince William Sound, Alaska. Owing to island biogeography and extreme northern latitude, forests arrived recently (less than 1,000 years ago) and coni-fers are limited to just Sitka spruce. The rest of the archipelago is treeless (al-though spruce is naturally migrating southward on Kodiak Island) and home to some of the most abundant salmon and brown bear populations in the world. Because of its remoteness and relatively high level of intactness on public lands, the Chugach National Forest is a conservation priority.

Chugach National Forest

Along the shoreline of Prince William Sound (see figure 2-1) is a naturally dis-continuous band of rainforest where three conifer species meet—Sitka spruce, western hemlock, and mountain hemlock. The Chugach includes an extensive coastline of more than 4,800 kilometers rimmed by tidewater glaciers. A mo-saic of rainforest, peatland bog, and beach fringe (an ecotonal zone where rain-forest and beachfront meet) mixes with meadow, shrubland, and tundra. The re-treat of glaciers allowed plants to colonize from Asia and regions north and west (Cook and MacDonald 2001; Cook et al. 2006).

The Chugach National Forest is the second largest national forest in the U.S. (2.24 million hectares), encompassing the Kenai Peninsula, Prince William Sound, and the Copper River Delta, one of the most productive migratory-shorebird stopovers in the world (Ricketts et al. 1999). About a third of the Na-tional Forest consists of active glaciers and nonforest habitat, including the largest coastal wetlands on the Pacific Coast. Most of the Chugach (85 percent of the forest base) is protected (see chapter 10) and the National Forest is managed primarily for fish, wildlife, and recreation so that threats are relatively minimized (although federal roadless-area policies are currently in flux; Turner 2009).

Industrial-scale logging and road building (primarily on Native Corporation lands and largely completed), mining, coastal transport of crude oil, energy development, and expanding recreation (mainly commercial tourism from large cruise ships and growing off-highway vehicle access) are principal threats. Nearly all (98 percent—2.16 million hectares) of the National Forest is roadless and, therefore, of international importance.

Currently, roadless-area policies in the United States are in flux: a roadless conservation rule excluding nearly all forms of logging and road building from all federally inventoried roadless areas (larger than 2,000 hectares) on national forests was enacted by President Clinton in 2001, repealed by President Bush in 2005, and partially reinstated by the courts in 2006 and 2008 with some notable exceptions (Tongass National Forest was exempted in 2003 by President Bush but not yet reinstated by court or administrative decisions), where it remains today in judicial and administrative uncertainty (although congressional legislation was recently introduced and the Obama administration has upheld most of the original rule thus far). In addition, ongoing effects to coastal species from the running aground of the Exxon Valdez in 1989 (at the time, the largest oil spill in U.S. history) are still being felt in some places (Short et al. 2007).

Conservation priorities for the Chugach include:

- Comprehensive planning for the growth of recreation-based tourism that is bringing both positive (economic diversification) and negative impacts (overuse) to the region.
- Tighter restrictions and enforcement of off-highway vehicle access that contributes to wildlife harassment, pollution, and soil erosion and compaction.
- Permanent protection of roadless areas, maintaining much of the pristine character of the rainforest.
- Full implementation of the Exxon Valdez oil-spill restoration plan, especially land acquisitions that compensate Native Corporations willing to sell their lands to the government so they can be protected from development (e.g., Afognak Island Native corporation lands, see www.cooperativeconservatonamerica.org).

PERHUMID RAINFOREST CONSERVATION PRIORITIES

These are wet places year round, with cool summers delivering 10–20 percent of the annual precipitation (Alaback and Pojar 1997). Snow is common but relatively transient near the coast. Conifer richness is higher than more northerly latitudes and includes Sitka spruce, western hemlock, mountain hemlock, amabilis fir, shore pine, western red cedar, and Alaska yellow cedar. High-priority conservation areas in this zone include the Tongass National Forest in southeast Alaska, the British Columbia Raincoast (also known as the Great Bear Rainforest and Haida Gwaii), and Clayoquot Sound (British Columbia). These areas were chosen because they support relatively high proportions of intact, old-growth rainforest, and abundant salmon and other wildlife, including apex carnivores.

Tongass National Forest

A vast chain of islands (including more than 5,500 larger than 0.4 hectares) and a narrow mainland coast stretch from the Yakutat Bay (north of Glacier Bay) to Dixon Entrance at the southern end of Prince of Wales Island, a distance of over 835 kilometers (Albert and Schoen 2007a; see figure 2-1). The "crown jewel" of the national forest system and a land that time nearly forgot, the Tongass is the largest (6.8 million hectares) national forest in the U.S. Verdant rainforest is spread across an extensive shoreline of nearly 30,000 kilometers, interspersed with tidewater glaciers and rugged islands, a profusion of salmon-spawning streams, and bisected by deeply dissected fjords (Albert and Schoen 2007a). About half (3.3 million hectares) of the total land base is forested but only about a third of the total land base (2.3 million hectares) is considered productive (i.e., commercially valuable) forest land; of that, about 90 percent (2 million hectares) is old growth (Albert and Schoen 2007a).

But not all old growth is the same in these rainforests. Stands of the largest old-growth trees (measured by total timber volume per hectare) have always been rare and today represent just 3 percent (~216,000 hectares) of the entire land base on the Tongass, because that is where much of the past logging has been concentrated (Albert and Schoen 2007a; USDA Forest Service 2008). These stands include some of the most valuable fish and wildlife habitats (DellaSala et al. 1996; Albert and Schoen 2007a) in the area. Past high-grade logging (focused on the largest trees) eliminated the best old growth concentrated on the most productive regions (especially Prince of Wales, Mitkof, and Chichagof islands) and the most ecologically valuable and intact watersheds (Albert and Schoen 2007a, b).

All six species of Pacific salmon—chum (*O. keta*), coho or silver (*O. kisutch*), king or chinook (*O. tshawytscha*), pink (*O. gorbuscha*), sockeye (*O. nerka*), and steelhead (*O. mykiss*)—spawn in Tongass waters, supporting some of the largest concentrations of brown bears and bald eagles in the world (Albert and Schoen 2007a). Fishing (commercial, subsistence, recreation) provides the greatest number of natural-resource jobs (Everest 2005). Many of the resources that people value are associated with intact watersheds having abundant old growth and healthy populations of fish and wildlife.

On the Tongass, logging and road building began in earnest after World War II, during which three 50-year government contracts (cancelled in the mid-1990s) were signed by the USDA Forest Service with private companies in Ketchikan, Wrangell, and Sitka to process Tongass lumber, mainly for export to Asia. These contracts were the first and only in the nation and focused logging efforts on high-volume rainforests with the largest trees (Sisk 2007b). A postwar logging boom began rainforest depletion that was slowed by litigation, poor export markets, and ultimately the loss of large blocks of easily accessible and economically valuable timber. Meanwhile, the Alaska National Interest Lands Conservation Act (ANILCA 1980) left an indelible mark on the Tongass landscape. One of ANILCA's legislative achievements was to protect some key places with productive fish and wildlife habitats such as Admiralty Island. However, much of the wilderness protection occurred in areas dominated by rock and ice that were not the most productive fish and wildlife habitats available (Albert and Schoen 2007b). Worst of all, ANILCA mandated large logging subsidies and unsustainable rates of logging that led to significant losses of old growth during the 1980s. The Tongass Timber Reform Act (1990) intended to eliminate these logging subsidies by reducing logging pressure on the high-volume (large-tree) rainforests and by protecting a number of invaluable old-growth watersheds, although abuses have been documented (Katz 1992).

Decades of rainforest logging on the Tongass have left many scars easily viewed in a flyover in one of Alaska's many float planes (a primary mode of Alaska transport, see plates 1a and 1b). This includes more than 7,900 kilometers of logging roads (USDA Forest Service 2008), especially on Prince of Wales Island (figure 2-2 shows the progression) and Long Island (over 90 percent of the forests were logged after they were transferred to Native corporations), the most biologically productive islands on the Tongass. In addition, whereas over half of the Tongass is without roads (3.74 million hectares), current national roadless-areas policies do not provide inviolate protection for rainforest (the region was exempted by a recent court ruling), with logging projects proposed for roadless areas by the Forest Service. Moreover, under the

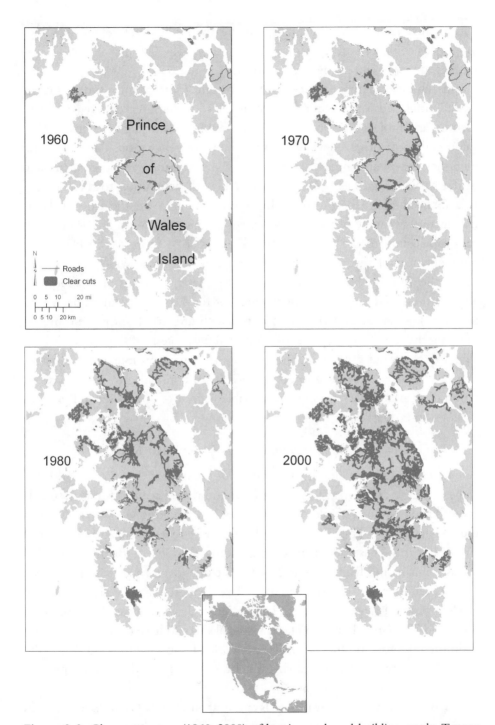

Figure 2-2. Chronosequence (1960–2000) of logging and road building on the Tongass National Forest, Prince of Wales Island, southeast Alaska. Source: Conservation Biology Institute and World Resources Institute.

current Tongass Land Management Plan, nearly 6,000 kilometers of new log-ging roads could be constructed and an additional 180,000 hectares of old for-est logged by the end of this century (USDA Forest Service 2008).

In response to these concerns, The Nature Conservancy and Audubon Alaska employed a technique known as conservation area design (CAD) to identify and propose for protection intact watersheds of highest conservation value and to prioritize developed watersheds for restoration. The CAD also identifies potential timber harvest areas where logging can be concentrated in the smallest land base within watersheds with preexisting road infrastruc-ture (Schoen and Albert 2007). This approach builds on the existing land-management strategy of the Tongass by adding intact watersheds with conser-vation priorities to the reserve network in order to more effectively ensure representation of diverse forest types in protected areas widely distributed across the Tongass. For instance, under the current Tongass land allocation, about a third of the forest's habitat values for focal species (salmon, deer, bear, murrelets), estuaries, and large-tree old growth are strictly protected. Imple-menting the CAD would place an additional 34 percent of the forest's habitat values in protected watersheds and another 15 percent in integrated manage-ment watersheds (a combination of protection for core old-growth areas, resto-ration of riparian areas, and some sustainable harvest of upland second growth) (Schoen and Albert 2007).

Great Bear Rainforest and Haida Gwaii (aka Raincoast)

As one of the few remaining large blocks of comparatively unmodified land-scapes on Earth, the Great Bear Rainforest and the adjacent offshore archipel-ago of Haida Gwaii are biologically rich, aesthetically unique, and rare (see fig-ure 2-1; plate 2a). This 7.4 million–hectare region encompasses the North Coast, Central Coast, and Haida Gwaii forest districts of British Columbia (Coast Information Team Report 2004). It includes over a quarter of the Pacific Coastal rainforests of North America (based on the rainforest distribution model, chapter 1) and some of the largest, relatively intact coastal rainforests in the world. The Great Bear and Haida Gwaii also could have been easily dubbed Rainforest Fjords, for the extensive network of mountainous fjords, or the Large Carnivore–Salmon Rainforest for the keystone role that salmon provide to apex predators like wolves and bears.

Like the Tongass, the Great Bear Rainforest and Haida Gwaii harbor one of the last opportunities for studying the outcome of long-term evolution on a geographic scale, and observing highly specialized and coevolved interactions that are being replaced elsewhere with invasive species or intensively managed landscapes (Paquet et al. 2006). Unfortunately, the rate of human-induced environmental change has been and may continue to be so rapid (Moola et al. 2004) that many species may not be able to keep pace with accelerating habitat losses.

Although the Raincoast has remained remarkably inaccessible when compared with other British Columbia regions, nevertheless extensive habitat loss and fragmentation have occurred in the southern portion of the Great Bear Rainforest and throughout much of central and northern Haida Gwaii. Even though most of the remaining forested land base is considered unsuitable for forestry (Pojar et al. 1999), many of the unprotected and biologically rich valley-bottom, old-growth forests have been logged or have been leased for logging to several major forestry companies and First Nations. Furthermore, new technologies such as helicopter logging provide access to old-growth timber once thought to be out of the reach of commercial interests (e.g., steep-forested slopes).

These magnificent forests are home to prodigious salmon runs (~2,500—Temple 2005), grizzly bears, wolves, Queen Charlotte goshawks (*Accipiter gentilis laingi*), and marbled murrelets (*Brachyramphus marmoratus*). Notably, a unique white phase of black bear, the Kermode bear (*U. americanus kermodei* or "Spirit Bear" (see plate 2b) because of its spiritual significance to First Nations, ranges from Roderick-Pooley and Princess Royal Islands in the south to the Nass Valley in the northern portion of the Great Bear Rainforest of British Columbia. Roughly 1 in 10 bears are born with a white coat (Ritland and Marshall 2001). Gribbell Island (20,600 hectares) on the British Columbia Central Coast has been identified in recent genetic studies as one of four important nearshore islands on the coast for this subspecies. Over 30 percent of the small bear population on the island are estimated to be white-phase bears, and this island is "richest in white bears" and has "substantial genetic isolation" (Ritland et al. 2001). Apparently, "Kermodism" has been established and maintained in populations by a combination of genetic isolation and somewhat reduced population sizes in insular habitat. Further, McCrory and Paquet (2008) consider the small insular population of between 90 and 130 bears to be vulnerable to ongoing logging and other human disturbances.

Fortunately, a recent land-use agreement between officials from the British Columbia government and local First Nations increased protection of forests from all industrial development in the Great Bear Rainforest portion of the Raincoast from ~9 percent to 28 percent (2.1 million hectares). An additional 5 percent of the land base (370,000 hectares) is off limits to logging but is not strictly protected from mining, road-building and other development. The remainder of the region's forests (so-called "matrix landscapes") are available to special forest management, called ecosystem-based management (EBM—see box 2-2) under which the amount of old-growth forest that can be logged across the landscape, in each watershed, and in each ecosystem type will be capped. Across the region as a whole a minimum of 50 percent of the natural level of old-growth forest of each ecosystem type will have to be maintained (or recruited over time in areas that have been heavily impacted already) under new EBM regulations. This translates to an additional 700,000 hectares of old-growth forest that will be off limits to logging (but not other development) outside of formally designated conservation areas in the region. Over the next

---

BOX 2-2

Ecosystem-Based Management of the Great Bear Rainforest: Solution or Incomplete Strategy?

The Great Bear Rainforest Agreement includes ecosystem-based management of industrial forests situated outside of the formally designated protected areas (i.e., the matrix landscape). These "EBM Zones" cover most (~70 percent) of the Great Bear Rainforest, including most of the timber harvesting landbase (e.g., valley bottoms and adjacent mid-slopes targeted by the logging industry—Martin et al. 2004), and the most biologically valuable areas in the region—areas of relative intactness and high aggregate conservation value (Rumsey et al. 2004). Though EBM was narrowly defined by an independent science panel for the region as a comprehensive suite of "best-practices" for forestry operations at multiple spatial scales (e.g., protection of rare and endangered plant communities, establishment of unlogged buffers adjacent to high-value fish habitat—see www.citbc.org), not all of the prescriptive elements recommended by the science panel have been legislated. Rather, EBM remains subject to further negotiations among First Nations, government, and other stakeholders, and is supposed to be fully implemented by 2014.

Nevertheless, there remains great uncertainty as to the ecological efficacy of newly established EBM planning and on-the-ground logging practices to maintain biodiversity on the landscape outside of formally designated protected areas. The lack of a representative reserve network necessitates establishment of a rigorous EBM process that eventually incorporates the 40–70 percent protection targets recommended by scientists through additional set-asides.

five years, government, environmental groups, industry, and First Nations have committed to revise the EBM regulations to increase the amount of old growth that will have to be maintained on the land base outside of protected areas (i.e., maintain 70 percent of natural levels of old growth over time). However, much uncertainty remains with the EBM process, which took 9 years to develop and another 5 years to implement, while logging continues. This means that considerable more old growth and biological diversity will have been lost by the time the agreements are finally implemented in 2014.

An accompanying recent land-use agreement with local First Nations in Haida Gwaii increased overall protection (strictest designation) on the islands to 0.5 million hectares. At least half of this archipelago is now protected and, as in the Great Bear Rainforest, new logging practices under EBM agreements will apply to forestry operations outside of established parks and protected areas.

Overall, the Great Bear Rainforest and Haida Gwaii agreements provide a significant increase in the overall area now strictly protected across the Raincoast (2.6 million hectares; 35 percent) from logging and all other industrial practices (e.g., mining) in parks, conservancies, and other legally designated areas. However, concerns have been raised over the inadequate size and spatial configuration of the protected areas network (Gilbert et al. 2004; Martin et al. 2004), incomplete representation of ecosystems (Wells et al. 2003), including coastal islands and productive old-growth stands (Paquet et al. 2004), inadequate levels of protection and omissions of key areas of critical habitat for focal species such as bears and wolves that were recommended by scientists (Gonzales et al. 2003; Martin et al. 2004; Paquet et al. 2004), and lack of full protection from all potentially deleterious human impacts (Gilbert et al. 2004; Moola et al. 2004). Perhaps more fundamentally, because much of the Great Bear Rainforest and Haida Gwaii are still relatively intact compared to other regions, it presents one of the last opportunities on the planet to conserve large, expansive wildlands (including old-growth rainforests) in their natural state.

For these and other reasons, both the Great Bear Rainforest and the Tongass are globally significant (Ricketts et al. 1999) in that the core elements necessary for conservation are mostly present and not in need of restoration at this time (e.g., predator-prey relationships are still intact—see Noss 2000). In particular, this region is especially deserving of core and comprehensive protection at the regional scale to ensure the persistence of these ecological values in perpetuity. For the current agreement to provide lasting conservation benefits, it needs to expand the size of protected areas that are presently still too small to maintain viable populations of wide-ranging species such as grizzly bears, coastal wolves, and salmon; to increase connectivity among existing and newly created protected areas; and to improve representation of rare, sensitive, and at-risk ecosystems such as islands and intact watersheds, along with rare, large old-growth trees.

Finally, ecosystem management and forestry operations throughout British Columbia should include both more responsible practices such as those of the Forest Stewardship Council (FSC),[1] and also protection of critical habitat of declining species (see box 2-3). Notably, about 1 million hectares of the EBM zones were recently certified under FSC standards, although it is unclear whether certification will make marked improvement to on-the-ground EBM measures at this time. Improved logging practices outside protected areas, in general, would reduce extirpation risks of vulnerable species (e.g., wolves, bears). But ultimately the reserve network needs to be expanded to include protection of large blocks (greater than 5,000 hectares) of intact and underrepresented watersheds, as these areas are crucial to salmon and other wildlife (e.g., large carnivores) and few intact watersheds this large remain outside Alaska and northern British Columbia (Beebe 1991). Given that the most rapid human-caused extinctions have occurred on islands worldwide (Pimm 1991), existing land-use plans for the Great Bear Rainforest and Haida Gwaii, like the Tongass, need to address fully the archipelago environment and associated extirpation risks.

Clayoquot Sound

The west coast of Vancouver Island (see figure 2-1) includes 265,000 hectares of mostly old-growth rainforest within the larger Clayoquot Sound region (3 million hectares). Vancouver Island has lost three-quarters of its productive old-growth forest (e.g., valley bottoms where the biggest trees typically grow) to logging, mining, agriculture, and urban development to date (www.viforest .org). Currently, ~35 percent (91,400 hectares) of the Sound's forest is under strict legal protection in parks and other protected areas (however, see table 10-1 for minor differences in protected-areas estimates based on GIS analysis). An additional 67,800 hectares within the Scientific Panel Watershed Reserves (EBM zones) is off limits to logging, but are still vulnerable to mining, roads, and other land-use activities. These areas were not included in the protected-areas database used in Chapter 10, as they were not strictly protected, but they do count toward assessments of regional protections.

---

[1]www.fsc-bc.org/britishcolumbia.htm

BOX 2-3

Endangered Birds of Southwest British Columbia: Going, Going, Gone?

The southwest Coast Mountains of British Columbia are home to the northern spotted owl, a forest raptor clinging to existence in the only place where it is found in Canada. An apex avian predator of older forests, fewer than a dozen owls remained in the wild in 2008, until the Canadian government removed most of them to establish a captive breeding program. Unfortunately, Canada's Minister of the Environment at the time denied habitat-protection measures mandated under Canada's endangered species law (i.e., the Species At Risk Act), deciding instead that "the northern spotted owl does not currently face imminent threats to its survival or recovery." This surprising decision was made as owl habitat continues to be logged, primarily by the government itself, and where over 80 percent of the old growth has already been logged since the 1940s.[a] Therefore, if conservation measures are not taken soon, the owl will be extirpated from its rainforest habitat within a wing beat of its evolutionary time line.

But the spotted owl is not the only bird in trouble. An unusual coastal seabird, the marbled murrelet, has declined by 70 percent over a 25-year period throughout Alaska and British Columbia (Piatt et al. 2006). This robin-sized seabird nests reclusively in clumps of moss in the tallest and oldest rainforest trees, returning infrequently to feed its young the catch-of-the day caught at sea. Cumulative impacts from climate-related changes in the marine ecosystem and human activities (e.g., logging of nest sites, forest fragmentation, gillnet bycatch, oil pollution) are suspected in the bird's decline (USFWS 1997). Immediate conservation is needed to stem further losses.

---

[a]See www.davidsuzuki.org/Forests/Canada/BC/Spotted_Owl.asp

The Sound's surroundings consist of highly dissected mountains blanketed by coastal rainforests and peppered by fjords and salmon-bearing rivers, although portions (e.g., the Upper Kennedy River) of this region are actually within the transitional rainforest type (Clayoquot Sound Scientific Panel 1995). Recognizing these distinctions, researchers subdivided the Sound into three biogeoclimatic units based on elevation, vegetation, and maritime influences. Within the units, old-growth rainforests are characterized by trees ranging from saplings to 1,000-year-old giants (Clayoquot Sound Scientific Panel 1995). Notably, most forest-dwelling species in this region nest in riparian (streamside) areas (72 percent), followed by older (greater than 140 years) forests (46 percent), and lastly by young, even-aged plantations (9 percent—Clayoquot Sound Scientific Panel 1995).

In 2000, Clayoquot Sound was designated a UNESCO Biosphere Reserve, though this was purely symbolic and did not result in any additional protected areas. Logging (under new EBM rules), fish-farming, exploratory drilling for a potential open-pit copper mine, and other ecologically damaging practices continue. Numerous open net-cage salmon farms now populate the Sound, as well as elsewhere in British Columbia. Open net-cage salmon farming has been linked to parasitic sea lice in juvenile salmon; toxic-waste effluent, including methylmercury contaminants, fish dyes, and growth hormones; and the release of antibiotics (Krkosek et al. 2005). Farmed Atlantic salmon kept in pens also are known to escape, where they genetically dilute wild Pacific stocks through hybridization.

To keep Clayoquot Sound's global status as a UNESCO Biosphere Reserve, we recommend that the British Columbia government legally protect remaining intact valleys, co-managed with First Nations, and provide funds in support of a conservation-based economy and restoration of degraded watersheds. The use of closed-containment facilities for farming salmon, combined with the elimination of exotic Atlantic salmon and improved waste water treatment, would dramatically elevate conditions for wild salmon in the Sound and throughout British Columbia in general.

SEASONAL RAINFOREST CONSERVATION PRIORITIES

Seasonal rainforests stretch from central Vancouver Island along the United States Pacific coastline to southern Oregon, with a secondary band farther inland along the western Cascades, a distance of some 604 kilometers (see figure 2-1). These rainforests are cloaked in fog and ocean spray, with annual precipitation ranging from ~2,000 to 4,080 millimeters. Contrary to the near year-round rain in the perhumid zone, most rainfall is distributed in the fall and winter months, with less than 10 percent in the dry summer (Alaback and Pojar 1997). Because of intermittent seasonal rainfall and the irregular but more frequent occurrence of fire (stands replaced every 90 to 250 years; Spies 2004), some researchers have elected not to classify them as temperate rainforest. However, we included them as well as the warm-temperate rainforests to the south because of their coastal association and similarities in plant and animal communities with their northern counterparts (also see Alaback and Pojar 1997). Conservation priorities in this zone include the Pacific Northwest, Siskiyou Wild Rivers (a subset of the Pacific Northwest Coast), and the Western Cascades (old forests).

Pacific Northwest

Pacific Northwest rainforests (4.8 million hectares) extend from the Olympic Peninsula to southern Oregon (see figure 2-1). They are ecologically more diverse than northern rainforests and are dominated by Douglas-fir, western hemlock, western red cedar, Sitka spruce, and shore pine (especially where salt spray is a factor), mixing with broadleaf hardwood understories in places. Some of world's largest hemlock and spruce are found on the Olympic Peninsula, where trees can exceed 2 meters in diameter and tower to over 60 meters (Kirk and Franklin 1992). Olympic rainforests are rich in plants from the ground up, with numerous ferns and shrubs blanketing the understory and unique epiphytes on the tops of the tallest trees (Kirk and Franklin 1992, plate 3).

Siskiyou Wild Rivers

At the terminus of the seasonal coastal rainforest in southwest Oregon is a distinct area known as Siskiyou Wild Rivers. This ~405,000-hectare area stands out as the "Pacific Coastal Outback" because it contains the largest complex of coastal roadless areas between the Mexican and Canadian borders. It supports one of the most prolific wild salmon runs south of Canada, exceptional concentrations of endemic plant species, large stretches of undeveloped and scenic rivers, and high levels of amphibian and avian richness (DellaSala et al. 1999). The western, coastal portion of this area is rainforest: plant communities farther inland are characteristic of drier, more fire-adapted forest types, although high-elevation areas retain rainforest characteristics due to increased moisture levels.

Western Cascades

This secondary rainforest band farther inland (see figure 2-1) includes coniferous forests dominated mainly by Douglas-fir, western hemlock, and western red cedar, and is distributed within three major physiographic provinces: the Northern Cascades, the Southern Washington Cascades, and the Western Cascades (Franklin and Dyrness 1973). Rainforest transitions to drier types at the Cascade Crest heading eastward toward drier continental conditions.

Old-growth rainforests have declined throughout each of these regions, from nearly two-thirds of the forest age classes historically to roughly 20

BOX 2-4

*Impacts of Logging and Road Building in Temperate and Boreal Rainforests
of the Pacific Coast of North America.*

Logging-related disturbances can have significant impacts on rainforest
biota, primarily from the mechanical disruption of understory plants and
forest soils, and from the oversimplification of the composition and structure of tree species, including the loss of coarse, woody debris and legacy
trees (old trees and structures carried over from the pre-disturbed forest).
Repeated logging on the same site every 30–100 years creates a long-
term "successional debt" whereby old-growth forest is depleted and thus
wildlife habitat and other ecosystem services degraded. Following logging, single species of trees (monocultures) are often planted in tight rows
(in the north, however, natural regeneration precludes the need for re-
planting), creating impoverished forests compared to the complex, multi-
species, multilayered rainforests they replaced.

The long-term adverse consequences of clearcut logging and road
building are many and include greater vulnerability of deer to loss of optimal winter habitat (Schoen et al. 1988) and human-caused wildlife mortality along roads, particularly for brown bears, wolves, and marten
(Schoen et al. 1994; Person et al. 1996; Trombulak and Frissell 2000; Person 2001; Flynn et al. 2004). Other terrestrial consequences include decline of wolves dependent on deer as a primary prey species, and increased
likelihood of conflicts between hunters and wolves for deer (Darimont
and Paquet 2002). Federally threatened species like the northern spotted
owl and marbled murrelet have declined precipitously due, in large part,
to loss of old-growth habitat caused by high rates of logging. Aquatic consequences include chronic degradation of salmon habitat by erosion and
siltation, increased water temperature due to loss of forest-related shading
along streams (see Reeves et al. 2006), and greater risk of mass-wasting
events (landslides) associated with road building and logging (Guthrie
2002).

Many plants and animals, particularly salmon, are now listed under the
U.S. Endangered Species Act and/or the Canadian Species at Risk Act,
due, in part, to logging, although legal listing has not necessarily led to effective protection of critical habitat (Yezerinac and Moola 2006). Consequently, resource managers need to consider long-term consequences of

---

BOX 2-4

Continued

forest management (including roads and logging techniques) on predator-prey dynamics and threatened species, recognizing that mitigation of those consequences may not always be possible. The archipelago landscape of coastal Alaska and British Columbia may amplify such impacts, given the greater vulnerability of island populations to extirpations (Hanley et al. 2004, Cook et al. 2006).

---

percent today (this estimate includes all temperate forests and not just rainforest—Strittholt et al. 2006). Notably, few old-growth rainforests remain along the coastline from southern Oregon to the Olympics, where staggered logging units have created a "checkerboard" pattern of alternating rainforest and clear-cuts, with much of the remaining unlogged rainforest restricted to federal lands. Logging, particularly of older forests, has impacted the area in many ways (see box 2-4).

The three regions are no strangers to logging conflicts, as made widely known by the 1990 listing of the northern spotted owl (*Strix occidentalis caurina*) as a federally threatened species (other species listings soon followed). A court injunction in 1993 shut down logging on federal lands, resulting in a compromise, known as the Northwest Forest Plan, which lowered allowable logging levels on federal lands by 80 percent while placing about a third of the region's public forests in late-successional reserves (LSRs). Since 1994, nearly 9.8 million hectares of federal lands have been managed under this plan, a global model in biodiversity protection and ecosystem management (DellaSala and Williams 2006). While the region is managed under various land-use categories, only about 5 percent is strictly protected by parks and wilderness areas (see chapter 10), although lower levels of protection are also provided on federal lands managed under the Northwest Forest Plan as late-successional reserves. There have been administrative attempts to weaken this reserve network, and logging has occurred in some reserves following wildland fire (DellaSala and Williams 2006).

Notably, the conservation foundation for the Northwest Forest Plan on federal lands is a state-of-the art reserve network developed by scientists (FEMAT 1993) to maintain the viability of old-growth associated species, primarily the northern spotted owl and marbled murrelet. Despite its significance as a global conservation model, only about 60 percent of the ~3 million hectares of reserves actually include old-growth forests (older than 150 years), with the rest made up of plantations that over time (50–100 years) may eventually acquire, through restorative silvicultural actions, older forest attributes (Strittholt et al. 2006). In addition, reserves are not inviolate, as forest thinning is permitted under certain conditions and post-disturbance logging has degraded reserves in places (e.g., Siskiyou Wild Rivers). Significant amounts of older forests (e.g., 0.4 million hectares of late-successional forest) remain in the "matrix" where logging is concentrated. In particular, logging on state and private lands is 3 to 4 times greater than on federal lands (Staus et al. 2002), making federal lands stick out as the last stronghold for old-growth rainforests and intact watersheds in this region. Invasive species, spread by logging operations, road building, livestock, and other dispersal agents, threaten to replace native species. Sudden Oak Death syndrome, caused by an introduced fungal species (*Phytophthora ramorum*), has crept into southern Oregon, where it threatens a host of trees, including coastal redwoods just to the south. Climate change may facilitate the spread of at least some of these deleterious invaders.

Conservation priorities for all three areas include:

- Protect remaining mature (100 years old) and old-growth forests on public lands; mature forests provide replacement trees for older ones that die from natural disturbances such as fire and wind throw.
- Prioritize intact watersheds and roadless areas (e.g., Siskiyou Wild Rivers) for protection as climate refugia for salmon, and other areas for restoration, particularly where road densities are high (e.g., the Oregon Coast Range).
- Implement invasive-species containment measures, particularly by curtailment of vectors that facilitate their spread (e.g., roads, livestock grazing, logging, and vehicles).
- Strengthen federal protection for threatened species, including northern spotted owl, marbled murrelet, and coho salmon, by increasing habitat protections for older forests and roadless areas.
- Thin forests by removing small trees in order to restore forest structure and composition in overly simplified tree plantations, and extend timber harvest rotations to grow older forests.
- Promote responsible forest management (such as FSC certification) on nonfederal lands to complement conservation strategies on federal lands.
- Designate new protected areas, particularly relatively intact areas spanning elevation gradients and important migration and travel corridors in order to allow species to find new habitat in response to climate change.

WARM RAINFOREST CONSERVATION PRIORITIES

Warm-temperate rainforests stretch in a long, narrow, and discontinuous band (skipping San Francisco Bay) no wider than 50 kilometers all the way from just above the Oregon-California border to just below the Bay area (see figure 2-1). This zone, which includes coastal redwood, is characterized by mild, wet winters and cool, dry summers on the coast with warmer dry summers inland. Less than 5 percent of the annual rainfall occurs during dry summer months, and snow is rare. Annual precipitation in the northern redwoods (Del Norte and Humboldt counties) exceeds 3,200 millimeters (Sawyer et al. 2000); it is generally drier southward.

Few trees on Earth are as impressive as coastal redwoods in tree size or age (see Sawyer et al. 2000 for trees over 100 meters tall). Only the *alerce* of Chile (see chapter 5), Huon pine (*Lagorastrobus franklinii*) of Tasmania (see chapter 8), and mountain ash (*Eucalyptus regnans*) of southeastern Australian (see chapter 8) approximate redwood trees in age or stature. Overall, the redwood region is globally significant (Ricketts et al. 1999) because it is the only place on Earth where coastal redwoods thrive, and because the area has higher species richness and more endemics than any of the Northern Hemisphere's rainforests (Alaback 1995; Noss et al. 2000).

A redwood's immense stature is enabled, in part, by its immersion in year-round fog that helps to meet moisture requirements, as well as by the biogeochemical properties and hydrological cycles of this rainforest (Sawyer et al. 2000). Interestingly, when redwood forests are cut down, fog production declines due to diminished evapotranspiration (Sawyer et al. 2000). Fire occurs infrequently (once every 250 to 500 years) in northern redwoods, more frequently inland and to the south (33 to 50 years), and more so in upland areas (less than 17 to 175 years—Sawyer et al. 2000). However, the more frequent fires are mostly ground based and this has allowed redwoods to reach enormous size and impressive longevity.

Redwood forests cover an estimated 647,000 (Sawyer et al. 2000) to 877,396 hectares, from extreme southwestern Oregon to southern Monterey County, California (35°N latitude).[2] Noss et al. (2000) subdivided the region into northern, central, and southern zones (with 25 additional subsections), owing to differences in plant assemblages, precipitation, and the influence of fire. In general, northern redwood forests more closely resemble rainforest counterparts to the north (in terms of conifer species composition), while central and southern redwood forests are a mixture of conifer and hardwood trees

---

[2]www.frap.cdf.ca.gov/data/frapgisdata/download.asp?rec=redwood/

(Sawyer et al. 2000). In addition to coastal redwood, these rainforests include white fir (*A. concolor*), sugar pine (*P. lambertiana*), incense cedar (*Calocedrus decurrens*), ponderosa pine (*P. ponderosa*), Sitka spruce, western red cedar, western hemlock, and a rich assortment of understory hardwoods, shrubs, ferns, and forbs (Sawyer et al. 2000).

The wildlife composition of redwood rainforest is similar to that of the transitional temperate rainforest to the north, with considerable overlap in species distributions (Cooperrider et al. 2000). Much of the richness of these forests is invisible to the untrained eye. At the top of the tallest trees, a diverse canopy ecosystem consists of unique invertebrates, epiphytes, mosses, lichens, small mammals, and even some salamanders that never venture from the rainforest canopy and some that never leave the same tree (see Noss et al. 2000). Likewise, the soil fauna are exceptionally rich, even more so than in tropical rainforests (Cooperrider et al. 2000).

Less than 4 percent of the redwood forests remain intact (Ricketts et al. 1999) and much of this is vulnerable to logging, as the region is mostly (83 percent) in private ownerships. Noss et al. (2000) indicate about 13 percent of the redwood region is strictly protected (but see table 10-1 for differences), and most of this is limited to just three areas in northern California—Humboldt Redwoods State Park, King Range National Conservation Area, and Redwoods National Park. Consequently, in the absence of additional conservation measures, old-growth redwood forests will remain restricted to a few isolated parks, important on many levels, but not in the form of intact ancient redwoods. (Much of the parks' forests are now recovering from logging prior to their protective designations.) Principal threats include the logging of remaining redwood groves (clusters of huge trees) mainly on nonfederal lands, exotic-species invasions (including Sudden Oak Death, which is starting to show up in redwood), and decline of salmon runs primarily from logging and road building (see Noss et al. 2000). Climate change brings uncertain prospects for these coastal giants, particularly as coastal fog has declined markedly in recent decades, as attributed to climate change (Johnstone and Dawson 2010).

Perhaps nowhere else in the coastal rainforest region is conservation more critical and urgent than in the redwoods. Without stepped-up conservation, the few remaining old-growth redwood forests will become outdoor museum pieces, reminders of a bygone era when the redwoods prospered. In response, Noss et al. (2000) introduced a three-part conservation strategy based on a CAD approach that recommends: (1) increased representation of various ecosystem types in strictly protected reserves for all three subregions; (2) conservation of focal species such as the Pacific fisher (*Martes pennanti pacifica*)—an old-growth-redwood meso-carnivore; and (3) protection of special elements such as "hot spots" of rare species, critical watersheds, and redwood groves. Building on the work of these scientists, the following conservation priorities for the redwood rainforest region are recommended:

- Establish new protected areas by purchasing redwood groves on private lands (for land acquisition priorities—www.savetheredwoods.org).
- Restore (regrow) degraded forests and watersheds as in several of the redwood parks.
- Greatly expand invasive-species and vector-containment measures, particularly through additional research on Sudden Oak Death.
- Maintain viable populations of focal species such as the Pacific fisher.
- Research the effects of climate change, including development of appropriate strategies that better enable redwood forests to adapt.

WHAT WILL IT TAKE TO SAVE PACIFIC COASTAL RAINFORESTS?

North America's coastal rainforests have stood the test of time against a backdrop of volcanic eruptions, retreating glaciers, wind storms, and occasional fires. As glaciers retreated, new areas opened up for colonizing species with ancient Beringia, continental, and coastal affinities. Today, tidal glaciers in the north are melting and logging has eliminated many of the region's giant trees. The naturally fragmented nature of the northern rainforest archipelago has served as a cradle of evolution, an outdoor laboratory for island biogeography, and a wake-up call alerting us to the vulnerabilities of rainforest island systems to human disturbances that often exceed the capacity of rainforest species to adapt.

It is the dawn of a new era for these remarkable rainforests, and their fate will be determined by whether prudent conservation measures are adopted by local, regional, and national governments in a time of accelerating global climate change. Both the U.S. and Canadian governments have demonstrated that enlightened conservation leadership can secure an enduring legacy for rainforest species and human communities. Although rainforest conservation will continue to be informed by the best available science of the times such as that provided by the scientific panels in British Columbia, Clayoquot Sound, and the Pacific Northwest, its mercurial nature depends on the will of the people, the politicians they elect, and the ongoing search for innovative and responsible solutions to humanity's ever-growing environmental footprint. Ultimately, these remarkable rainforests and the human societies around them are joined at the hip, "like grizzly bear to salmon and forest to rain." If we allow the remaining intact areas and the vulnerable species in them to drift toward oblivion, humanity will suffer irreplaceable losses, including the inability to experience rainforest giants that sprang from tiny seedlings when indigenous people and the earliest explorers paid homage to their splendor.

North America's coastal rainforests are indeed deserving of stepped-up attention if they are to persist in these challenging times. This is urgently needed in order to ensure that the rainforest giants in coastal redwood forests and older forests of the region, rich early successional rainforests, roadless watersheds, and the apex carnivore–salmon food web continues to flourish. Humanity will increasingly depend on the myriad ecosystem services these rainforests have provided through the ages. But these rainforests will only continue to nurture us if they are protected and properly managed.

LITERATURE CITED

Alaback, P. B. 1982. Dynamics of understory biomass in Sitka spruce–western hemlock forest of southeast Alaska. *Ecology* 63:1932–48.

———. 1984. Plant succession following logging in the Sitka spruce–western hemlock forests of southeast Alaska: Implications for management. USDA Forest Service General Technical Report PNW-173.

———, and G. P. Juday. 1989. Structure and composition of low-elevation old-growth forests in research natural areas of southeast Alaska. *Natural Areas Journal* 9 (1):27–39.

———. 1995. Biodiversity patterns in relation to climate: The coastal temperate rainforests of North America. Pp. 105–33 in *High-latitude rain forests and associated ecosystems of the west coast of the Americas: Climate, hydrology, ecology and conservation*. Ed. R. Lawford, P. Alaback, and E. R. Fuentes. *Ecological Studies* 116. Berlin: Springer-Verlag.

———, and J. Pojar. 1997. Vegetation from ridgetop to seashore. Pp. 69–88 in *The rainforests of home*. Ed. P. K. Schoonmaker, B. von Hagen, and E. C. Wolf. Washington, DC: Island Press.

Albert, D., and J. Schoen. 2007a. A conservation assessment for the coastal forests and mountains ecoregion of southeastern Alaska and the Tongass National Forest. Chap. 2 in *The coastal forests and mountain ecoregion of southeastern Alaska and the Tongass National Forest*. Ed. J. Schoen and E. Dovichin. Anchorage, AK: Audubon Alaska and The Nature Conservancy. www.conserveonline.org/workspaces/akcfm

———. 2007b. A comparison of relative biological value, habitat vulnerability, and cumulative ecological risk among biogeographic provinces in southeastern Alaska. Chap. 3 in *The coastal forests and mountain ecoregion of southeastern Alaska and the Tongass National*

Forest. Ed. J. Schoen and E. Dovichin. Anchorage, AK: Audubon Alaska and The Nature Conservancy. www.conserveonline.org/workspaces/akcfm

Allendorf, F. W., and R. S. Waples. 1996. Conservation and genetics of salmonid fishes. Pp. 238–80 in *Conservation genetics: Case histories from nature.* Ed. J. C. Avise and J. L. Hamrick. Toronto: Chapman and Hall.

Augerot, X. 2004. *Atlas of Pacific salmon: The first mapped-based status assessment of salmon in the North Pacific.* Portland, OR: Ecotrust.

Beebe, S. B. 1991. Conservation in temperate and tropical rain forests: The search for an ecosystem approach to sustainability. Pp. 595–603 in *Transactions of the 56th North American Wildlife and Natural Resources Conference.*

Carstensen, R., J. Schoen, and D. Albert. 2007. Overview of the biogeographic provinces of southeastern Alaska. Chap. 4 in *The coastal forests and mountain ecoregion of southeastern Alaska and the Tongass National Forest.* Ed. J. Schoen and E. Dovichin. Anchorage, AK: Audubon Alaska and The Nature Conservancy. www.conserveonline.org/workspaces/ akcfm

Cederholm, C. J., D. H. Johnson, R. E. Bilby, L. G. Dominguez, A. M. Garrett, W. H. Graeber, E. L. Greda, et al. 2000. Pacific salmon and wildlife: Ecological contexts, relationships, and implications for management. Pp. 628–85 in *Wildlife-habitat relationships in Oregon and Washington.* Ed. D. H. Johnson and T. A. O'Neil. Olympia, WA: Washington Department of Fish and Wildlife.

Clayoquot Sound Scientific Panel. 1995. Sustainable ecosystem management in Clayoquot Sound. Report 5, Chaps. 1–2: Introduction / The Clayoquot Sound Environment: Hishuk ish ts'awalk. www.cortex.ca/dow-cla.html

Coast Information Team Report. 2004. December 28, 2004 revised report. www.citbc .org

Cook, J., and S. MacDonald. 2001. Should endemism be a focus of conservation efforts along the North Pacific Coast of North America? *Biological Conservation* 97:207–13.

———, N. G. Dawson, and S. O. MacDonald. 2006. Conservation of highly fragmented systems: The north temperate Alexander Archipelago. *Biological Conservation* 133:1–15.

Cooperrider, A., R. F. Noss, H. H. Welsh Jr., C. Carroll, W. Zielinski, D. Olson, S. K. Nelson, and B. G. Marcot. 2000. Terrestrial fauna of redwood forests. Pp. 119–64 in *The redwood forest: History, ecology, and conservation of the coast redwoods.* Ed. R. F. Noss. Washington, DC: Island Press.

Darimont, C. T., and P. C. Paquet. 2002. The gray wolves, *Canis lupus,* of British Columbia's central and north coast: distribution and conservation assessment. *Canadian Field-Naturalist* 116:416–22. www.people.ucsc.edu/~darimont/publications/Darimont

———, T. E. Reimchen, and P. C. Paquet. 2003. Foraging behavior by gray wolves on salmon streams in coastal British Columbia. *Canadian Journal of Zoology* 81:349–53.

———, P. C. Paquet, and T. E. Reimchen. 2008. Spawning salmon disrupt tight trophic coupling between wolves and ungulate prey in coastal British Columbia. *BMC Ecology* 8:14.

DellaSala, D. A., J. C. Hagar, K. A. Engel, W. C. McComb, R. L. Fairbanks, and E. G. Campbell. 1996. Effects of silvicultural modifications of temperate rainforest on breeding and wintering bird communities, Prince of Wales Island, southeast Alaska. *Condor* 98:706–21.

———, S. B. Reid, T. J. Frest, J. R. Strittholt, and D. M. Olson. 1999. A global perspective on the biodiversity of the Klamath-Siskiyou ecoregion. *Natural Areas Journal* 19:300–319.

———, and J. Williams. 2006. Northwest Forest Plan ten years later—how far have we come and where are we going? *Conservation Biology* 20:274–76.

Everest, F. H. 2005. Setting the stage for the development of a science-based Tongass land management plan. *Landscape and Urban Planning* 72:13–24.

Fedje, D. W., and R. W. Mathewes, eds. 2005. *Haida Gwaii: Human history and environment from the time of the loon to the time of the iron people.* Vancouver: Univ. of British Columbia Press.

FEMAT (Forest Ecosystem Management Assessment Team). 1993. Forest ecosystem management: An ecological, economic, and social assessment. Report of the FEMAT. U.S. Washington, DC: Government Printing Office.

Flynn, R., T. Schumacher, and M. Ben-David. 2004. Abundance, prey availability and diets of American martens: Implications for the design of old-growth reserves in southeast Alaska. U.S. Fish and Wildlife Service final report. Alaska Department of Fish and Game.

Franklin, J. F., and C. T. Dyrness. 1973. *Natural vegetation of Oregon and Washington.* Corvallis, OR: Oregon State Univ. Press.

———, T. A. Spies, R. Van Pelt, A. Carey, D. Thornburgh, D. R. Berg, D. Lindenmayer, M. Harmon, W. Keeton, and D. C. Shaw. 2002. Disturbances and structural development of natural forest ecosystems with silvicultural implications, using Douglas-fir forests as an example. *Forest Ecology and Management* 155:309–423.

———, and J. Agee. 2003. Scientific issues and national forest fire policy: Forging a science-based national forest fire policy. *Issues in Science and Technology* 20:59–66.

Fujimori, T. 1971. Primary production of a young *Tsuga heterophylla* stand and some speculations about biomass of forest communities on the Oregon coast. USDA Forest Service Research Paper PNW-123.

Gallant, A. L. 1996. USGS ecoregions of Alaska map. www.explorenorth.com/library/maps/ecoreg-alaska.html

Gavin, D. G., L. B. Brubacker, and K. P. Lertzman. 2003. Holocene fire history of a coastal temperate rainforest based on soil charcoal radiocarbon dates. *Ecology* 84:186–201.

Gende, S. M., R. T. Edwards, M. F. Willson, and M. S. Wipfli. 2002. Pacific salmon in aquatic and terrestrial ecosystems. *BioScience* 52:917–28.

Gilbert, B., L. Craighead, B. Horejsi, P. Paquet, and W. P. McCrory. 2004. Scientific criteria for evaluation and establishment of grizzly bear management areas in British Columbia. Panel of independent scientists. Victoria, BC. www.raincoast.org

Gonzales, E. K., P. Arcese, R. Schulz, and F. L. Bunnell. 2003. Strategic reserve design in the central coast of British Columbia: Integrating ecological and industrial goals. *Canadian Journal of Forest Research* 33:2139–50.

Goward, T., and A. Arsenault. 2001. Cyanolichens and conifers: Implications for global conservation. *Forest, Snow and Landscape Research* 75:303–18.

Gresh, T., J. Lichatowich, and P. Schoonmaker. 2000. Salmon decline creates nutrient deficit in Northwest streams. *Fisheries* 15 (1), January 2000:15–21.

Guthrie, R. H. 2002. The effects of logging on the frequency and distribution of landslides in three watersheds on Vancouver Island, British Columbia. *Geomorphology* 43:273–92.

Halpern, C. B., D. McKenzie, S. A. Evans, and D. A. Maguire. 2005. Initial responses of forest understories to varying levels and patterns of green-tree retention. *Ecological Applications* 15:175–95.

Hanley, T. A., W. P. Smith, and S. M. Gende. 2004. Maintaining wildlife habitat in southeastern Alaska: Implications of new knowledge for forest management and research. *Landscape and Urban Planning* 72:113–33.

Hebda, R. J., and C. Whitlock. 1997. Environmental history. Pp. 227–54 in *The rainforests of home*. Ed. P. K. Schoonmaker, B. von Hagen, and E. C. Wolf. Washington, DC: Island Press.

Johnstone, J. A., and T. A. Dawson. 2010. Climatic context and ecological implications of summer fog decline in the coast redwood region. PNAS early edition. www.pnas.org/cgi/doi/10.1073/pnas.0915062107

Katz, D. 1992. Tongass at the crossroads: Forest service mismanagement in the wake of the Tongass Timber Reform Act. AFSEE white paper to the Clinton administration. Association of Forest Service Employees for Environmental Ethics. Eugene, OR.

Keith, H., B. G. Mackey, and D. B. Lindenmayer. 2009. Re-evaluation of forest biomass carbon stocks and lessons from the world's most carbon-dense forests. *Proceedings of the National Academy of Sciences* 106 (28):11635–40.

Kirk, R., and J. Franklin. 1992. *The Olympic rainforest: An ecological web*. Seattle: Univ. of Washington Press.

Krkosek, M., M. A. Lewis, and J. P. Volpe. 2005. Transmission dynamics of parasitic sea lice from farm to wild salmon. *Proceedings of the Royal Society of London* B, March 2005.

Lertzman, K, T. Spies, and F. Swanson. 1996. From ecosystem dynamics to ecosystem management. Pp. 361–82 in *The rainforests of home*. Ed. P. K. Schoonmaker, B. von Hagen, and E. C. Wolf. Washington, DC: Island Press.

Lichatowich, J. 1999. *Salmon without rivers*. Washington, DC: Island Press.

Lindenmayer, D. B., and J. F. Franklin. 2002. *Conserving forest biodiversity: A comprehensive multiscaled approach*. Washington, DC: Island Press.

Luyssaert, S., E. Detlef Schulze, A. Börner, A. Knohl, D. Hessenmöller, B. E. Law, P. Ciais, and J. Grace. 2008. Old-growth forests as global carbon sinks. *Nature* 455:213–15.

Martin, D., F. M. Moola, B. Wareham, J. Calof, C. Burda, and P. Grames. 2004. Canada's Rainforests: 2004 Status Report. David Suzuki Foundation, Vancouver. www.canadianrainforests.org

Montgomery, D. R. 1997. The influence of geological processes on ecological systems. Pp. 43–68 in *The rainforests of home*. Ed. P. K. Schoonmaker, B. von Hagen, and E. C. Wolf. Washington, DC: Island Press.

Moola, F., D. Martin, B. Wareham, J. Calof, C. Burda, and P. Grames. 2004. The coastal temperate rainforests of Canada: The need for ecosystem-based management. *Biodiversity* 9 (5):9–15.

Mote, P.W., A. F. Hamlet, M. P. Clark, and D. P. Lettenmaier. 2005. Declining mountain snow-pack in western North America. *American Meteorological Society,* January 2005:39–49.

Muñoz-Fuentes, V., C.T., Darimont, R. K. Wayne, P. C. Paquet, and J. A. Leonard. 2009. Ecological factors drive differentiation in wolves from British Columbia. *Journal of Biogeography* 31:1516–31.

Nehlsen, W., and J. A. Lichatowich. 1997. Pacific salmon: Life histories, diversity, productivity. Pp. 213–26 in *The rainforests of home.* Ed. P.K. Schoonmaker, B. von Hagen, and E. C. Wolf. Washington D.C.: Island Press.

Noss, R. F., J. R. Strittholt, G. E. Heilman Jr., P. A. Frost, and M. Sorensen. 2000. Conservation planning in the redwoods region. Pp. 201–28 in *The redwood forest.* Ed. R. F. Noss. Washington, DC: Island Press.

———. 2000. Maintaining the ecological integrity of landscapes and ecoregions. Pp. 191–208 in *Ecological integrity: Integrating environment, conservation and health.* Ed. D. Pimentel, L. Westra, and R. F. Noss. Washington, DC: Island Press.

Nowacki, G. J., and M. G. Kramer. 1998. The effects of wind disturbance on temperate rainforest structure and dynamics of southeast Alasaka. USDA Forest Service General Technical Report. PNW-GTR-421.

———, P. Spencer, T. Brook, M. Fleming, and T. Jorgenson. 2001. Ecoregions of Alaska and neighboring territories. www.agdc.usgs.gov/data/projects/fhm

Paquet, P.C., C. T. Darimont, R. J. Nelson, and K. Bennett. 2004. A critical examination of protection for key wildlife and salmon habitats under proposed British Columbia Central Coast Land and Resource Management Plan. Raincoast Conservation Society, Victoria, BC. www.raincoast.org

———, S. M. Alexander, P. L. Swan, and C. T. Darimont. 2006. Connectivity conservation. Pp. 130–56 in *Influence of natural landscape fragmentation and resource availability on distribution and connectivity of marine gray wolf (Canis lupus) populations on Central Coast, British Columbia, Canada.* Ed. K. R. Crooks and M. Sanjayan. New York: Cambridge Univ. Press.

Person, D. M. Kirchhoff, V. Van Ballenberghe, G. Iverson, and E. Grossman. 1996. The Alexander Archipelago wolf: A conservation assessment. General Technical Report, PNW-GTR-384. U.S. Forest Service.

———. 2001. Alexander Archipelago wolves: Ecology and population viability in a disturbed insular landscape. Doctoral Dissertation, Univ. of Alaska, Fairbanks.

Piatt, J. F., K. J. Kuletz, A. E. Burger, S. A. Hatch, V. L. Friesen, T. P. Birt, M. L. Arimitsu, G. S. Drew, A. M. A. Harding, and K. S. Bixler. 2006. Status review of the marbled murrelet (*Brachyramphus marmoratus*) in Alaska and British Columbia. U.S. Geological Survey Open-File Report 2006-1387. www.pubs.usgs.gov/of/2006/1387/

Pojar, J., C. Rowan, A. MacKinnon, D. Coates, and P. LePage. 1999. Silivicultural options in the Central Coast. Report prepared for the Central Coast land and coastal resource management plan. www.ilmbwww.gov.bc.ca/citbc/b-SilviOpt-Pojar-Dec99.pdf

Price, M. H. H., C. T. Darimont, N. F. Temple, and S. M. MacDuffee. 2008. Ghost runs: Management and status assessment of Pacific salmon returning to British Columbia's central and north coast. *Canadian Journal of Fish and Aquatic Science* 65:2712–18.

Pimm, S. L. 1991. *The balance of nature?* Chicago: Univ. of Chicago Press.

Redmond, K., and G. Taylor. 1997. Climate of the coastal temperate rain forest. Pp. 25–42 in *The rainforests of home*. Ed. P. K. Schoonmaker, B. von Hagen, and E. C. Wolf. Washington, DC: Island Press.

Reeves, G., J. E. Williams, K. M. Burnett, and K. Gallo. 2006. The aquatic conservation strategy of the Northwest Forest Plan. *Conservation Biology* 20:319–29.

Reimchen, T. E. 2000. Some ecological and evolutionary aspects of bear-salmon interactions in coastal British Columbia. *Canadian Journal of Zoology* 78:448–57.

Ricketts, T., E. Dinerstein, D. Olson, C. Loucks, W. Eichbaum, D. DellaSala, K. Kavanagh, et al. 1999. *A conservation assessment of the terrestrial ecoregions of North America*. Washington, DC: Island Press.

Ritland, K., C. Newton, and D. Marshall. 2001. Inheritance and population structure of the white-phased "Kermode" black bear. *Current Biology* 11:1468–72.

Rumsey, C., J. Adron, K. Ciruna, T. Curtis, Z. Ferdana, T.D. Hamilton, K. Heinemeyer, et al. 2004. An ecosystem spatial analysis for Haida Gwaii, Central Coast, and North Coast British Columbia. Coast Information Team, Victoria, BC. www.citbc.org

Sawyer, J. O., S. C. Sillett, W. J. Libby, T. E. Dawson, J. H. Popenoe, D. L. Largent, R. Van Pelt, et al. 2000. Redwood trees, communities, and ecosystems: A closer look. Pp. 81–118 in *The redwood forest: History, ecology, and conservation of the coast redwoods*. Ed. R. F. Noss. Washington, DC: Island Press.

Schoen, J., M. Kirchhoff, and J. Hughes. 1988. Wildlife and old-growth forests in southeastern Alaska. *Natural Areas Journal* 8:138–45.

———, R. Flynn, L. Suring, K. Titus, and L. Beier. 1994. Habitat capability model for brown bear in Southeast Alaska. *International Conference on Bear Research and Management* 9:327–37.

———, and D. Albert. 2007. Southeastern Alaska conservation strategy: A conceptual approach. Chap. 10 in *The coastal forests and mountain ecoregion of southeastern Alaska and the Tongass National Forest: A conservation assessment and resource synthesis*. Ed. J. Schoen and E. Dovichin. Audubon Alaska and The Nature Conservancy, Anchorage, AK. www.conserveonline.org/workspaces/akcfm

Schofield, W. B. 1988. Bryogeography and the bryophytic characterization of biogeoclimatic zones of British Columbia, Canada. *Canadian Journal of Botany* 66:2673–86.

Short, J. W., G. V. Irvine, D. H. Mann, J. M. Maselko, J. J. Pella, M. R. Lindeberg, J. R. Payne, W. B. Driskell, and S. D. Rice. 2007. Slightly weathered Exxon Valdez oil persists in Gulf of Alaska beach sediments after 16 years. *Environmental Science Technology* 41 (4):1245–50.

Sillett, S. C. 1999. Tree crown structure and vascular epiphyte distribution in *Sequoia sempervirens* rain forest canopies. *Selbyana* 20:76–97.

Simenstad, C. A., M. Dethier, C. Levings, and D. Hay. 1997. The terrestrial/marine ecotone. Pp. 149–88 in *The rainforests of home*. Ed. P. K. Schoonmaker, B. von Hagen, and E. C. Wolf. Washington, DC: Island Press.

Sisk, J. 2007a. The southeastern Alaska salmon industry: Historical overview and current status. Chap. 8 in *The coastal forests and mountain ecoregion of southeastern Alaska and the Tongass National Forest*. Ed. J. Schoen and E. Dovichin. Audubon Alaska and The Nature Conservancy, Anchorage, AK. www.conserveonline.org/workspaces/akcfm

————. 2007b. The southeastern Alaska timber industry: Historical overview and current status. Chap. 9 in *The coastal forests and mountain ecoregion of southeastern Alaska and the Tongass National Forest.* Ed. J. Schoen and E. Dovichin. Audubon Alaska and The Nature Conservancy, Anchorage, AK. www.conserveonline.org/workspaces/akcfm

Smith, W. P. 2004. Evolutionary diversity and ecology of endemic small mammals of southeastern Alaska with implications for land management planning. *Landscape and Urban Planning* 72:135–55.

Smithwick, E. A. H., M. E. Harmon, S. M. Remillard, S. A. Acker, and J. F. Franklin. 2002. Potential upper bounds of carbon stores in forests of the Pacific Northwest. *Ecological Applications* 12:1303–17.

Spies, T. A. 2004. Ecological concepts and diversity of old-growth forests. *Journal of Forestry* 103:14–20.

Staus, N. L., J. R. Strittholt, D. A. DellaSala, and R. Robinson. 2002. Rate and pattern of forest disturbance in the Klamath-Siskiyou ecoregion, USA. *Landscape Ecology* 17:455–70.

Strittholt, J. R., D. A. DellaSala, and H. Jiang. 2006. Status of mature and old-growth forests in the Pacific Northwest, USA. *Conservation Biology* 20:363–74.

Swanson, M. E., J. F. Franklin, R. Beschta, C. Crisafulli, D. A. DellaSala, R. L. Hutto, D. B. Lindenmayer, and F. J. Swanson. 2010. The forgotten stage of forest succession: Early-seral successional ecosystems on forest sites. *Frontiers in Ecology and Environment.* doi:10.1890/090157

Temple, N., ed. 2005. *Salmon in the Great Bear Rainforest and Haida Gwaii.* Raincoast Conservation Society, Victoria, BC.

Turner, T. 2009. *Roadless rules: The struggle for the last wild forests.* Washington, DC: Island Press.

Trombulak, S., and C. Frissell. 2000. Review of ecological effects of roads on terrestrial and aquatic communities. *Conservation Biology* 14:18–30.

USDA Forest Service. 2008. Tongass land management plan, Final Environmental Impact Statement. Alaska Region, Ketchikan, AK.

USFWS (U.S. Fish and Wildlife Service). 1997. Recovery plan for the threatened marbled murrelet (Washington, Oregon, and California populations). USFWS Region 1, Portland, OR.

VanPelt, M. 2008. *Identifying old trees and forests in western Washington.* Washington State Department of Natural Resources, Olympia, WA.

Veblen, T. T., and P. B. Alaback. 1996. A comparative review of forest dynamics and disturbance in the temperate rainforests in North and South America. Pp. 173–213 in *High-latitude rain forests and associated ecosystems of the west coast of the Americas: Climate, hydrology, ecology and conservation.* Ed. R. Lawford P. Alaback, and E. R. Fuentes. *Ecological Studies*116. Springer-Verlag.

Wells, R. W., F. L. Bunnell, D. Hagg, and G. Sutherland. 2003. Evaluating ecological representation within differing planning objectives for the central coast of British Columbia. *Canadian Journal of Forest Research* 33:2141–50.

Wimberly, M. C., T. A. Spies, C. J. Long, and C. Whitlock. 2000. Simulating historical variability in the amount of old forests in the Oregon Coast Range. *Conservation Biology* 14:167–80.

Western Regional Climate Center. 2006. Historical monthly precipitation levels. www
.wrcc.dri.edu/htmlfiles/ak/ak.ppt.html

Willson, M. E., and K. C. Halupka. 1995. Anadromous fish as keystone species in vertebrate
communities. *Conservation Biology* 9:489–97.

Worley, I. A. 1972. The bryogeography of Southeastern Alaska. Ph.D dissertation. Univ. of
British Columbia, Vancouver, BC.

Yezerinac, S. M., and F. Moola. 2006. Conservation status and threats to species associated
with spotted owls: A new flagship fleet for British Columbia. *Biodiversity* 6:3–9.

CHAPTER 3

Temperate and Boreal Rainforests of Inland Northwestern North America

Dominick A. DellaSala, Paul Alaback, Lance Craighead,
Trevor Goward, Paul Paquet, and Toby Spribille

On the windward slopes of the Columbia and Rocky Mountains is a 7 million–hectare disjunct rainforest (see figure 3-1) that arguably includes the largest expanse of inland temperate and boreal rainforests on Earth (but also see Inland Southern Siberia, chapter 9). Most rainforests here are temperate except at the most northerly latitudes, where they grade to boreal rainforest. Although early biogeographers noted the unexpected presence here of numerous species typical of coastal regions (e.g., Daubenmire 1943), only recently have ecologists recognized these forests as a distinct entity: an inland counterpart to coastal rainforests of the Pacific Northwest (Alaback et al. 2000; Goward and Arsenault 2000; Goward and Spribille 2005).

RAINFOREST VITALS AND GLOBAL ACCOLADES

Inland rainforests are distributed discontinuously from latitudes 54°N to approximately 46°N in northern Idaho and extreme northwestern Montana (see figure 3-1). In the north, contemporary rainforest structure and composition appear to have developed some 6,000 to 2,000 years ago (Gavin et al. 2009), long after the retreat of the Pleistocene glaciers. Consequently, like their Pacific Coast counterparts (see chapter 2), much of the contemporary inland rainforest has existed in its present form for but a few generations of the oldest rainforest trees (Hebda and Mathewes 1984). Indeed, Gavin et al. (2009) postulate that

Figure 3-1. Temperate and boreal rainforest of Inland Northwestern North America (adapted from Craighead and Cross 2007), showing strictly protected areas (cross-hatching). Note the lack of overlap between rainforest and protected areas.

some of the oldest stands in northern portions of the inland rainforest may represent "first generation" forests as yet unaffected by wildfire subsequent to the establishment of western red cedar (*Thuja plicata*) and western hemlock (*Tsuga heterophylla*). In the south, rainforests escaped extensive glaciation, presumably providing an important refuge during the Pleistocene for moisture-seeking plants, mosses, hepatics (bryophytes), and especially lichens.

Lodgepole pine (*Pinus contorta*) was probably the first tree species to colonize this region after deglaciation (Hebda 1995); however, the oldest tree in terms of physical age is certainly red cedar. Red cedar is also usually the dominant tree in the region's oldest stands, some of which have grown undisturbed for over 1,000 years (Sanborn et al. 2006, plate 4a). Such forests are effectively multigenerational, older than the oldest trees within them; technically they are referred to as "antique" (Goward and Pojar 1998).

In this region, black bears (*Ursus americanus*) and grizzly bears (*Ursus arctos*) graze on forbs in the abundant lush avalanche chutes and, in winter, mountain caribou (*Rangifer tarandus*) forage on long tresses of lichens hanging from the conifer canopy in higher elevation, parkland forests near tree line in the so-called "snow forests" (Spribille 2002). Gray wolf (*Canis lupus*), grizzly bear, black bear, cougar (*Puma concolor*), lynx (*F. lynx*), and wolverine (*Gulo gulo*) make up the full complement of large carnivores (see table 3-1).

Viewed regionally, inland rainforests constitute a major anomaly, having closer biological affinities with distant coastal regions than with the dominant,

Table 3-1. Unique attributes of temperate and boreal rainforests of Inland Northwestern North America.

Attribute	Importance
Geographically restricted and distributed discontinuously along the western slopes of the Columbia and Rocky Mountains, with isolated pockets in the northwestern United States	Globally rare inland rainforest
Influenced by both continental climate (cool, dormant seasons with high snowfall, long winters, pronounced seasons) and coastal climate (Pacific storms)	Unique climatic conditions allow rainforest communities to persist
High overlap with Pacific Coastal forests in epiphytic lichens, mosses, fungi, ferns, and other vascular plants with coastal affinities	Important biotic feature that helps define these forests as rainforests
Low frequency of stand-replacing fire and other large natural disturbances where forests reach very old ages—"antique forest"	Globally rare forest-age classes
Major center of plant diversity, including diverse canopy epiphytes and some of the highest estimates of plant richness in coniferous forests of this latitude	Distinguishing biotic attribute of temperate rainforests, important in food-web dynamics (e.g., caribou)
Last remaining watersheds in the contiguous United States with full complement of large carnivores	Important in food-web dynamics (e.g., predator-prey) and globally rare

drought-tolerant Douglas-fir (*Pseudotsuga menziesii*), western larch (*Larix occidentalis*), and ponderosa pine (*P. ponderosa*) forests that surround them (Alaback et al. 2000). Indeed, many trees, shrubs, and ferns characteristic of the coastal rainforests also occur in the interior rainforest (Daubenmire 1943; Lorain 1988). Recent studies on lichens show an even stronger relationship between interior and coastal forests, especially in northern regions (Goward and Spribille 2005).

Although inland rainforests bear many similarities to the coastal rainforests of North America (see below), from a regional and global perspective they are unique (see table 3-1). They include some of the highest estimated lichen richness in temperate coniferous forests of this latitude, including many restricted species (Goward and Spribille 2005). Because these forests form in wet inclusions in a generally dry woodland region, they provide habitat for species requiring abundant soil moisture in the growing season (Daubenmire 1968). Also, as noted above, many vascular plants appear to be disjunct with coastal affinities following geological and geographic changes in the region, both before and during the late-Pleistocene glaciation some 10,000 years ago (Lorain 1988). Such forests are transitional between more typical dry Rocky Mountain forests, grasslands, and the boreal forests to the north (Daubenmire 1943; Alaback 2000).

Inland rainforests also share many of the same bird species, most of the same mammal species, and virtually all of the same trees. Importantly, they host a remarkable number of the "rainforest lichens" usually associated with the Pacific Coast (Goward and Spribille 2005). Many of these coastal species probably first got established here in the spray zones of waterfalls, only later moving out into the moist old forests. The role of waterfalls as colonization "toeholds" in small rainforest regions is little studied but potentially key to understanding population establishment in fragmented rainforest landscapes (Goward and Björk 2009). High richness of canopy epiphytes, and high bryophyte richness (Newmaster et al. 2003), make these rainforests extraordinarily important nodes of biodiversity among forests at this latitude (see table 3-1).

RAINFOREST CLASSIFICATIONS

Inland rainforests occur mostly at low elevations. At the northern edge of their occurrence in British Columbia their upper limits occur at about 1,000 meters elevation, but in the south they extend upward to 1,800 meters elevation. Portions of inland British Columbia, sufficiently moist to support cedar and hemlock on mesic sites, are often popularly referred to collectively as the "interior wet-belt" or, more technically, as the Interior Cedar-Hemlock (ICH) Zone (Ministry of Forests and Range 2008; see fig. 3-1). At higher elevations, the ICH zone gives way to the Engelmann Spruce–Subalpine Fir Zone (ESSF), with its namesake trees *Picea engelmanii* and *Abies lasiocarpa*, respectively. The lower boundary of the ESSF signals a transition from rainforests to "snow forests" (Spribille 2002). Understory floristics between the ICH and the ESSF zones, however, are very similar (Spribille 2002), possibly a function of insufficient time since deglaciation for full development of more characteristic sets of herbs and shrubs. Even so, these are highly distinct forest types characterized by cool snowy (ESSF), and, except in winter, warm, rainy (ICH) climates. In this chapter, we consider mainly the inland rainforest portion of the ICH zone except where important rainforest species like caribou overlap with snow forests.

In general, moist forests are relatively isolated in the Rocky Mountain region and are variously known as cedar-hemlock, interior cedar-hemlock or cedar-fir (*Abies lasiocarpa*) forests. Different from their coastal counterparts, the inland rainforests are not continuous; rather, they occupy isolated pockets of anomalously wet conditions that have formed on the windward slopes of several mountain ranges lying between the Coast Mountains on the west and the Rocky Mountains on the east. These ranges include the Cariboos, Monashees, Purcells, and Selkirks in Canada, with small outliers in the Cabinet and Bitterroot, Clearwater, St. Joe, and Coeur d'Alene Mountains farther south. A few disjunct inland rainforests also have formed on the western foothills of the Rocky Mountains themselves (Daubenmire 1969; Arsenault and Goward 2000).

RAINFOREST CLIMATE

The climate of the inland rainforest region is a mixture of maritime and continental influences that vary seasonally and regionally. For much of the year, precipitation is principally derived from easterly moving maritime air masses intersecting with the first of several longitudinally oriented mountain ranges rising east of the Columbia Basin, Okanagan Trench, and Cariboo-Chilcotin Basin of interior British Columbia. During the summer months, this pattern shifts in the southern region to more continental influences, with prevailing winds from Canada and the Great Plains. Average June rainfall for some southern portions of the inland rainforest can be greater than that for many coastal stations at similar latitudes (Alaback et al. 2000). Average rainfall in July and August, however, is substantially below even rain shadow–affected sites in near-coastal areas such as the Willamette Valley of Oregon or Puget Sound of Washington. Northward, the summer climate is cool and wet, like the Pacific Coast.

In keeping with their occurrence many hundreds of kilometers from the open ocean, inland rainforests are subject to annual temperature fluctuations nearly double those experienced in coastal rainforest at similar latitudes (see figure 1–3b in chapter 1). This is true especially in the north, where the annual march of temperature is decidedly "continental," similar in fact to that of boreal forests (Alaback et al. 2000). Nevertheless, any potential drying effect of these colder winters and warmer summers is offset by snowmelt from the winter snowpack coupled with relatively high early-summer precipitation. The relatively cool summer temperatures in the north are significant in increasing water runoff due to low evaporation rates. Cool July temperatures are in fact useful in distinguishing most temperate and boreal rainforest climates from surrounding drier forest types (e.g., Alaback 1991).

Despite the occasional high and extreme afternoon summer temperatures (~30°C) and long periods between significant rainfalls during the warmest months, even the most southerly interior rainforests support significant inland rainforest communities. Studies of soil moisture suggest that there is little consistent drought on these southerly inland sites (Daubenmire 1968). At a local level, the oldest stands are most often positioned along valley bottoms or concave slopes at the base of large mountainous terrain where moisture readily collects (Goward and Pojar 1998).

INLAND RAINFOREST DISTURBANCE DYNAMICS

Much like the rainforests of the Pacific Coast of North America, Chile, and Tasmania (Alaback 1991; Lawford et al. 1996), inland rainforests have a continuous but layered canopy of trees and shrubs from the ground up. The forest canopy is occasionally punctuated by the death of a rainforest elder, as light-seeking plants, many of which are the offspring of the dead tree, move in to fill the gaps.

Fires generally occur at intervals of about 80 to over 200 years (Smith and Fischer 1997), and up to 1,000 years or more in northern sections (Gavin et al. 2009). In the southern rainforest limit in Idaho and Montana, fire plays a key role in the structural appearance of rainforests. Pure western red cedar groves, for example, are often a product of frequent ground fires. Fires also typically limit the maximum age of hemlock-cedar forests in the south. Old-growth (or antique) rainforest pockets persist mostly in moist, valley-bottom microsites out of the reach of most fires. The prevalence of these forests increases with latitude as fires become less frequent, mostly due to increases in summer rainfall (Goward and Spribille 2005).

Other disturbances are also important in the ecology of these rainforests. Insect outbreaks, for example, are well known as a key natural disturbance agent in moist and subalpine forests in the region (e.g., Swetnam and Lynch 1993). Drought appears to amplify interacting disturbance agents such as fire and insects. Wind too is an important agent generally, and, in particular, in older moist or subalpine forests, but has been little studied (e.g. Alaback 2000).

OCEANIC LICHENS AS RAINFOREST INDICATORS

Lichens are dual entities: part fungi, part algae. As such, they require living conditions optimized not for one but two organisms. Likewise, the inability of lichens to produce "roots" capable of reaching subsurface water means they are not buffered (as flowering plants are) against rapid change. Taken together, the lichen "lifestyle" makes lichens a fine-tuned and continuously updated "barometer" of climate, nutrient supply, and environmental discontinuity. Lichens offer a convenient and easy-to-implement method of circumscribing environmental conditions such as those that make inland rainforests what they are. In addition, many lichens are character species of the oldest of inland rainforests (Goward 1994), being either sparse or absent from younger forest types (Radies and Coxson 2004).

Goward and Spribille (2005) used a set of typically oceanic "macrolichens" (relatively large leafy and shrubby lichen species commonly studied by ecologists) to map the extent of inland rainforests. They concluded that the distribution area of various tree-dwelling lichens, such as peppered moon (*Sticta fuliginosa*) and netted specklebelly (*Pseudocyphellaria anomala*), corresponds with the boundaries of inland rainforest, being restricted to wet-belt valleys from the Nakusp area of British Columbia north to the northern Robson Valley near Prince George. Further, this distribution type correlates with the absence of a summer moisture deficit, suggesting that it is physiological extremes, not averages, that limit the occurrence of these rainforest species. Most portions of the southern wet-belt are too dry in summer to support most such rainforest lichens. On the other hand, they do support many mid-coastal rainforest species not present farther north.

REGIONAL "CRADLE" OF LICHEN EVOLUTION

Strictly speaking, lichens are not a single evolutionary group; rather, the lichen "lifestyle" evolved multiple times independently in different groups of fungi. The inland rainforest supports a large cross-section of this diversity in the forest canopy and forest floor. Old-growth forests especially host many species restricted to the region (endemics) and others apparently unknown to science. Recently, Spribille et al. (2009) reported eight previously undescribed lichen species from inland rainforests, most occurring in antique forests in the "core" areas outlined by Goward and Spribille (2005). Many additional species still remain to be described. Some characteristic lichens of the inland rainforest appear to have close relationships to groups of lichens otherwise known only from the Valdivian and/or Tasmanian rainforests of the Southern Hemisphere. Included here are cryptic paw lichen (*Nephroma occultum*), pink dimple lichen (*Gyalectaria diluta*, nearest relatives in Valdivia and Tasmania), and brown bead lichen (*Schaereria brunnea*, related to Tasmanian species). Moreover, many endemic and relict species are concentrated in the humid cedar-hemlock forests of north-central Idaho, especially the Clearwater River drainage.

LARGE CARNIVORES OF THE INTERIOR WET-BELT

Most of the interior wet-belt, particularly in British Columbia, is very mountainous and remote. Particularly in the north, this is a land where people rarely settled, and farming and ranching are rare. Unlike drier areas where valleys are covered with rural homesteads scattered between towns, in the northern wet-belt, valley bottoms are relatively free of human settlement. Here, wolves, cougars, wolverines, coyotes (*C. latrans*), lynx, grizzly, and black bears are usually free to roam without being attracted to a hen house, a hobby-farmer's small flock of sheep, or fruit trees—historically (and often still) death sentences for large carnivores.

Both black and grizzly bears are abundant in most of the wet-belt ecosystems. There are ample spring foods in the numerous avalanche paths and alpine meadows, and in the summer, huckleberries are often abundant in open burns and the numerous regenerating cut-blocks. Grizzly bears are relatively abundant in the north, but their distribution and densities decline southward toward the international border. They are extirpated in the southern portion of the interior wet-belt in central Idaho. Wolverines appear to be relatively abundant in wet-belt ecosystems. Perhaps their abundance is a temporary phenomenon: a result of the high numbers of marmots (*Marmota* spp.) in the mountains in the summer and mountain goats (*Oreamnos americanus*), moose, and deer that die in winter.

MOUNTAIN CARIBOU AS A "FLAGSHIP" RAINFOREST SPECIES

The inland rainforests—and snow forests—of western North America are home to the mountain caribou (plate 4b), a southern outlier of the great caribou herds of the far north. In the past these animals roamed old-growth stands down to valley elevations, though nowadays they are pretty much restricted to the high country: the upper parkland forests and alpine (Edwards 1954). Only in early winter and again in early spring do they venture to lower elevations, in part forced down by hunger. In the long, snowy winter months in between, the fate of the mountain caribou hangs largely by a slender thread—a thread of hair lichens.

Caribou—and reindeer too—are unique among the ungulates in their reliance on lichens as a primary source of winter food. In most regions supporting caribou, winter snows are relatively shallow, and caribou are able to paw through the snowpack to feed on ground-dwelling lichens, especially reindeer lichens (*Cladina*). Conditions, however, are very different in North America's western interior mountains. Here, the winter snows can accumulate to tremendous depths—2 to 5 meters—placing the ground vegetation well out of reach. To survive, the mountain caribou must forage instead on tree-dwelling hair lichens—mostly horsehair lichen (*Bryoria*), but also witch's hair (*Alectoria*)—hanging in long tresses well above the snowpack. Without hair lichens, there would be no mountain caribou (Rominger et al. 1996); and without old-growth forests, there would be few hair lichens to support mountain caribou during the long, cold, and snowy mountain winter (Edwards 1954).

In ecology, an ultimate cause provides a general underlying reason for a phenomenon of interest, typically involving evolutionary and adaptive mechanisms, whereas a proximate cause is the immediate cause behind a phenomenon. Both ultimate and proximate causes regulate caribou numbers in the following ways.

Ultimate Causes

The ultimate reason why the mountain caribou are endangered is that their old-growth habitat was logged, flooded, burned, and fragmented into a landscape that can neither feed them nor provide the security they need to survive.

The relentless destruction and transformation of the region's older forests has deprived mountain caribou of their life requisites, while exposing them to levels of predation they did not evolve with and are incapable of adapting to.

Most mountain caribou now live in (and above) the wettest portions of the inland rainforest. Historically, other herbivore species—moose (*Alces alces*), deer (*Odocoileus* sp.), and elk (*Cervus elaphus*)—are believed to have occurred in low numbers in these areas. Few ungulates meant few predators; and this, combined with the widespread occurrence of old-growth forests throughout the rainforest region, resulted in a relative abundance of mountain caribou (Spalding 2000). The situation appears to have changed with the advent of warmer climatic conditions at the close of the Little Ice Age in the late 1880s. An increase in wildfire around this time initiated a decline in the amount of old-growth forest and a trend to younger forest types. This trend was accelerated considerably in the early 1900s by forest fires started—often deliberately—by prospectors, settlers, and, in some areas, railroaders. More recently the amount of old growth has declined further, owing to industrial-scale logging dating to the early 1970s and continuing today (see below). Logging has created vast tracts of young forests that are connected and dissected by networks of roads and trails. Just how widespread the resulting young, predator-rich forests have become can scarcely be appreciated except in an over-flight of interior British Columbia—these days easily managed (virtually) through GoogleEarth.

Proximate Causes

One result of this dramatic increase in young forests over the past century has been an equally dramatic increase in the numbers of moose and deer now inhabiting the inland rainforest region. Both of these kinds of animals benefit from the copious forbs and shrubs that are associated with young (or early seral) forests in this region (see Edwards 1954), so in comparison with historical populations both now occur here in relative "superabundance." Although impossible to quantify or verify, early reports suggest that the number of wolves, cougars, black bears, and coyotes also increased during this period, likely due to newly abundant prey (Edwards 1954). In theory, the buildup of early seral herbivores enabled wolf populations in the north and cougar populations in the south to attain levels much higher than would have existed before logging occurred.

In the winter, caribou live mostly high up in the mountain refugia, safely separated from the other large mammals in the valleys. But in summer, moose and deer move to higher elevations, and so do their predators. Moreover, industrial forestry, roads, and compaction and removal of snow for recreational and industrial purposes have been shown to influence winter wolf movements and predatory behavior, resulting in increased kill rates of caribou and other prey in previously secure winter ranges (Paquet et al., forthcoming). The combination of increased mortality due to predation with liberal hunting regulations in many areas appears to have greatly reduced the number of mountain caribou by the 1940s, particularly in the southern half of the wet-belt, and farther north by 1970.

The current trend toward increasingly extreme climatic conditions—deep snows one winter, shallow snows the next—is now forcing caribou periodically to leave the safety of high mountain winter strongholds for more marginal wintering habitat at lower elevations (Kinley et al. 2007). In winter, mountain caribou forage almost exclusively in habitats supporting tree-dwelling hair lichens in great abundance (Rominger et al. 1996). As already noted, abundant lichen populations are restricted to forests older than about 100 years (Edwards et al. 1960; Goward 1998, 2003); and it is here, in the old growth, that mountain caribou pass most of each winter. Yet recent forest practices in British Columbia have reduced much of this former habitat, while at the same time severely fragmenting the rest (see below). From this it seems to follow—though the hypothesis has not been tested—that predators *per se* are only one proximate cause of the ongoing caribou decline (see box 3-1); another, even more proximate cause may be the caribou's day-to-day search for hair lichens in sufficient quantity. It must be asked whether the search for food across an increasingly fragmented landscape is now forcing mountain caribou to spend more time in open country, hence putting themselves at risk of predators, poachers, and disturbance by snowmobilers.

THE MOUNTAIN CARIBOU'S CONSERVATION BURDEN

Protection efforts in the inland rainforest of British Columbia have a long history aimed at the mountain caribou. This effort began in earnest with the southward expansion of Wells Gray Provincial Park in the mid-1950s, mostly to protect the area's large but dwindling mountain caribou herds. Unfortunately, 70 percent of the old-growth mountain caribou habitat in Wells Gray Park was burned in several fires in 1926 and succeeding years (Edwards 1954). Other large parks were created—Glacier National Park, the Purcell Wilderness Conservancy, Valhalla Provincial Park—but these contained only small amounts of old-growth ICH. Very large areas include rock, ice, alpine meadows, and subalpine parkland. Northern forest districts in the inland rainforest region began

---

BOX 3-1

Large Carnivores vs. Mountain Caribou: Conservation Paradox?

The interior wet-belt represents a special challenge to conservation biologists concerned about the fate of charismatic species and ecosystems. Creating young forests (plantations) through logging ultimately triggered the situation today where caribou are threatened by predation from large carnivores that are otherwise considered flagship species in their own right. Thus, an increase in the number and size of protected areas is justifiable in order to slow the conversion of older forests to plantations for the sake of caribou and other rainforest species. This challenging situation, however, has created a sort of conservation paradox pitting large-carnivore biologists, who are used to large-carnivore scarcity in more southerly regions, against caribou biologists who believe predation to be a proximal threat to caribou (see Wilson 2009). While abundant populations of large carnivores are often cheered by conservationists, this issue has fueled disagreement over predator management in Canada. While the debate over predator-prey management lives on, protected areas remain the touchstone of virtually all national and international conservation efforts (Dudley 2008). Most conservation biologists have called for expanding them even in places where large carnivores are abundant, as the uncertainties of climate change and ongoing land-clearing activities outside protected areas are likely to magnify risks and challenges to conserving rainforest species.

---

setting aside old-growth forest for mountain caribou in the 1980s. Not surprisingly, these northern areas have the most caribou remaining.

In 1994, regional planning and implementation of land-use planning began to tackle conservation issues with the advent of the Commission on Resources and Environment (CORE). The mission of the CORE public processes was to create land-use plans that would double the protected areas to 12 percent of the land base (based on general recommendations of the Brundtland Commission 1987—see chapter 10 for further discussion), and to bring about forestry reforms that would constitute "sustainable development." The land-use plans were created through interest-based negotiations amongst public, industrial, and government stakeholders. These CORE processes evolved into land and resource management plans that were based on trade-offs between various interest groups. Forest conservation and protection of endangered species were only two of many interests. At the time, it was assumed that logging could take place in a way that preserved enough of the caribou's old-growth habitat through modified harvest zones which, despite leaving more and larger uncut blocks, have fragmented formerly intact areas.

The most significant achievement of these planning processes, in terms of protection of inland rainforest, was the creation of the Cariboo Mountains Provincial Park, which linked Wells Gray and Bowron Provincial Parks to create a large protected-area complex in the heart of the inland rainforest. Unfortunately, compromises required drawing the boundaries of the new protected area to include only very small amounts of the timber-harvesting land base. New protected areas in the Central Selkirk region—the Goat Range Provincial Park, the addition to the Purcell Wilderness Conservancy, the West Arm Provincial Park—all had 30–50 percent of their area excluded (old-growth forest/mountain caribou habitat) when the new parks were approved by the government. Important areas likewise were poorly represented, including the south Selkirks, the Robson Valley in the far northern end of the region, and the Revelstoke area, which contain some of the largest mountain caribou herds. In contrast, vast drainages such as the Kuskanax, Duncan, Lardeau, and Adams River watersheds were extensively logged at low and middle elevations, leaving only small patches and strips of inland rainforest. The net effect of these actions likely contributed to precipitous declines in caribou in just five years: down from approximately 2,450 mountain caribou in 1997 to about 1,900 animals in 2002 (Mountain Caribou Technical Advisory Committee 2002). A decade ago there were 17 isolated subpopulations in British Columbia. Now only 15 remain and many contain fewer than 20 animals. Clearly, caribou are sliding toward extinction, primarily because of logging of their primary habitat—the lichen-rich old-growth rainforest.

Recently the British Columbia government issued Government Action Regulations (GARs) for 380,000 hectares that include modified logging zones for caribou. However, these areas are mostly in high elevations, they omit a significant portion of very wet– and wet-forest types, and these putative "protected areas" are to remain in place provided caribou numbers show marked increases. They are also subject to numerous exemptions that could be granted under the discretion of the Regional Manager of the Minister of Environment (GAR Orders, Appendix I available from the Minister). Any future decline in the number of mountain caribou could result in these areas being once again opened up to industrial logging.

BOX 3-2
Forest Protection in Inland Northwestern North America:
What Should Count?

Conservation groups, governments, and local stakeholders often debate not only whether to set aside a particular region but what kind of human activities should be permitted within protected areas and how much a protected area or overall conservation strategy actually contributes toward overall conservation. For instance, the current area of old-growth rainforests is below historic levels (see table 3-2), and the conservation strategy for the region is dependent on caribou persistence and enforcement of forestry regulations in the Interior Cedar-Hemlock portion of British Columbia. Old-Growth Management Areas for caribou are generally distributed in a sea of clear-cuts, with the management areas comprised of large parcels that may include a combination of both suitable habitat areas (such as old Interior Cedar-Hemlock forests) and other non-habitat types (such as clear-cuts originating before protection was officially designated). This is why it is important that the type of protection and assumptions (regarding degree of protection and quality) made in conservation strategies be clearly defined by always following well-established approaches to protected-area inventories such as those developed by the IUCN (Dudley 2008) or the GAP analysis project (Scott et al. 1993). Because of uncertainties of government regulations, the British Columbia strategy would likely receive an IUCN land use code of VI—corresponding to protected areas managed mainly for the sustainable use of natural ecosystems but not in the strictest sense. Further, exemptions to the regulations can be permitted according to the discretion of the Regional Manager of the Minister of Environment.

Despite the Province's strong commitment to the sustained killing of predators, based on the assumption this would increase caribou numbers (Mountain Caribou Science Team 2005), the efficacy of caribou-management approaches remain controversial and questionable (see box 3-2). Wildlife science is equivocal concerning the long-term effects on ecosystems of widespread predator control (Orians et al. 1997). In fact, the removal of apex predators (especially wolves and cougars) can have profound and adverse ecological consequences, including disruption of established trophic relationships (Terborgh et al. 1999; Hebblewhite et al. 2005; Beschta and Ripple 2009). If caribou mortalities can be reduced by managing recreation and logging activities instead of by killing predators, then decision makers will need to decide whether these activities will continue to have primacy over the conservation requirements of caribou. If these factors continue to be mismanaged and the caribou disappear, then the old-growth forests may, once again, be on the chopping block.

HOW MUCH IS STRICTLY PROTECTED AND WHERE?

Throughout this book, we discuss the importance of protected areas in conservation strategies both globally and regionally. As conservation scientists, however, we are also concerned about whether the distribution of protected areas adequately represents high-conservation-value areas such as old-growth and intact rainforests important to caribou and a broad suite of species (e.g., many lichens). We now take a closer look at the distribution of protected areas in British Columbia, where we have good data, in order to help address the question of how well or poorly they represent high-conservation-value areas of the region. We chose strictly protected as our land-use category because such areas are legally protected and managed long-term for their scientific and wilderness values or for ecosystem protection and restoration (i.e., IUCN-protected areas, categories I and II—Dudley 2008). In addition, these particular categories best meet the IUCN definition of a protected area: "a clearly defined geographic space, recognized, dedicated, and managed, through legal or other effective means, to achieve the long-term conservation of nature with associated ecosystem services and cultural values" (Dudley 2008).

In order to best assess representation goals for rainforest protection, we grouped the analysis around two central questions:

(1) What proportion of the protected-areas network includes high-conservation-value areas such as old forests and intact areas?
(2) How representative are protected areas of specific forest subtypes, slopes, and elevation zones important to species like caribou?

Mapping assessments of protected areas for the British Columbia portion of the region were available for this analysis (personal communication, B. Cross, GIS Analyst at Applied Conservation GIS) and were based on biogeoclimatic zones, a classification system created by scientists and the British Columbia

Ministry of Forests principally to manage forests of the Canadian provinces (Ministry of Forests and Range 2008). Biogeoclimatic zones are defined as a geographic area having similar patterns of energy flow, vegetation, and soils as a result of a broadly homogenous macroclimate.

High-Conservation-Value Areas

Historically, over three-quarters of the inland rainforest region was forested (see table 3-2). However, human settlement has removed over 1.5 million hectares from the original forest base, and logging has degraded another 2 million hectares (~21 percent of the current forest land base). Remaining forest cover, as a percentage of the total land base, is nearly equally distributed as young- (26 percent) and old-growth forest (25 percent). Notably, these data were obtained from satellite images in 2000, with most of the analysis completed in

Table 3-2. Land-cover types and protected areas (parks, eco reserves >1,000 ha, IUCN I and II) in the inland temperate and boreal rainforests of British Columbia as of 2003.[a]

Land Cover Type	Area (ha)	Percent[b]
Total land base	14,311,387	
Rock and ice	786,709	5.5
Other non-forest	2,401,369	16.8
Original forested area	11,123,309	77.7
Human settlement	1,534,354	10.7
Current forest land base[c]	9,588,955	67.0
Young forest	3,755,131	26.2
Old forest	3,554,065	24.8
Logged forest	2,000,212	14.0
Roads and utility	222,719	1.6
Intact forest (>1,000 ha)	4,294,040	30.0
Intact forest (>5,000 ha)	3,244,975	22.7
Fragmented forest	5,294,915	37.0
Intact old forest	2,414,344	16.9
Intact old Interior Cedar-Hemlock forest	495,465	3.5
Protected forest (>1,000 ha)	2,444,310	17.1
Protected lands (non-forest)	1,104,667	7.7
Protected old forest	741,172	5.2
Protected intact forest	946,764	6.6
Protected intact old forest	641,948	4.5

[a]Data provided by Valhalla Wilderness Society, C. Pettitt, and B. Cross (personal communication).
[b]Percentages are based on comparisons to total land base. Due to overlap in cover types, percentages do not sum to a total of 100 percent.
[c]Based on the difference between estimated original forest cover and the amount of forest remaining after human settlement.

2003 (personal communication, C. Pettitt, Valhalla Wilderness Society). Although rate-of-change estimates for forests are lacking, logging continues at the expense of intact old forests (17 percent) and intact interior cedar-hemlock/wet forests (3.5 percent), with forest fragmentation (5.3 million hectares, 37 percent of total land base) already widespread. Further, despite their conservation importance, only a fraction of old (5 percent) and old, intact (4.5 percent) forests are strictly protected and thus most such forests are vulnerable to additional logging. For instance, nearly the entire Timber Harvesting Land Base (primarily older forests) in the range of the mountain caribou remains open to logging. Consequently, logging is replacing critical spring and early-winter habitat for mountain caribou with early successional habitat for super-abundant species like moose and deer.

Forest Subtypes, Elevation, and Slope

Based on biogeoclimatic classifications, ICH forests were grouped by three sub-types (very wet/cool, wet/cool, and moist). These subtypes correspond to differences in moisture levels and represent ecological differences in plant communities. ICH forests also were stratified by elevation and slope in order to further assess their representation in protected areas (see table 3-3). In general, the wet type had the highest (20 percent) representation in protected areas and the moist type the lowest (~8 percent). There were differences in protected areas based on slope, with wet forests having the highest representation on gentle-to-moderate slopes (0–40 percent slopes, combined ~16 percent) and the moist, steep type the lowest (>40 percent slope, 2.1 percent combined). ICH subtypes were best represented (~20 percent) in protected areas at mid elevations (1,000–1,500 meters), particularly the wet type; however, representation levels were low across all elevations, particularly the low and upper areas. Notably, high-elevation areas (above 1,700 meters) on moderate slopes (16–45 percent) generally receive high use by caribou, as do "warm" south- and west-(136–315°) facing aspects and forests over 250 years old (Stronen et al. 2007).

In sum, the British Columbia strategy, while a step forward in caribou and rainforest conservation, could be markedly improved if officials take three critical steps: (1) tighten the loopholes (exemptions) in the GARs and make them legally binding (i.e., enact lasting protection); (2) pass a strong species- and ecosystems-protection act that can be applied effectively to imperiled species like caribou on provincial lands (most public or Crown lands are under provincial jurisdictions not subject to federal protections such as afforded under the Species At Risk Act except under special circumstances; see Moola et al. 2007);

Table 3-3. Amount in hectares (with percentage of total area in protection shown in parentheses) of Interior Cedar Hemlock (ICH) by subtype (very wet, wet, moist),[a] slope, and elevation in inland northwestern British Columbia as of 2008.[b]

Slope (%)	Very Wet	Wet	Moist
0–20	15,219 (3.6)	91,608 (8.9)	70,080 (3.5)
20–40	15,893 (3.7)	71,602 (6.9)	51,091 (2.6)
40–60	14,780 (3.5)	42,363 (4.1)	33,604 (1.7)
>60	5,183 (1.2)	6,728 (0.6)	8,520 (0.4)
Protected Subtotals	51,075	(11.9)	212,301 (20.5)
163,295 (8.2)			
Elevation (m)			
0–1000	13,107 (3.1)	73,519 (7.1)	66,769 (3.4)
1000–1500	35,659 (8.3)	136,113 (13.2)	85,240 (4.3)
>1500	2,309 (0.5)	2,669 (0.3)	11,286 (0.6)
Protected Subtotals	51,075	(11.9)	212,301 (20.5)
163,295 (8.2)			
Total Area[c]	427,760	1,033,419	1,983,337

[a]Based on biogeoclimatic zones of British Columbia (Ministry of Forests and Range 2008). Only IUCN categories I and II (strict protection—Dudley 2008) were included.
[b]GIS analysis provided by B. Cross.
[c]Based on the sum of protected and unprotected lands. Totals were used to calculate protection percentages of the various cover types in the table.

and (3) greatly increase representation of high-conservation-value areas and subtypes across elevations and on gentle-to-moderate slopes.

CONSERVATION-AREA DESIGN IN THE UNITED STATES

Beginning in the 1990s, the World Wildlife Fund's Rocky Mountain Carnivore Project, which spanned both the United States and Canada, recognized a portion of the inland rainforest within a larger area of interest (Hackman and Paquet 1994). Some results of this plan were then incorporated into The Nature Conservancy Canadian Rocky Mountain Ecoregional Plan, completed in 2003. At about the same time that WWF was launching its carnivore strategy, the Yellowstone to Yukon (commonly called Y2Y) Conservation Initiative was born (www.y2y.net/). Y2Y was built on WWF's Rocky Mountain Carnivore Project, which stressed the importance of large-scale linkage zones for wolves and grizzly bears in the Yukon, British Columbia, Alberta, Idaho, Montana,

Wyoming, and Utah. The initiative mushroomed into an informal coalition of grassroots conservation organizations united in their vision of a network of protected core areas connected by wildlife movement corridors stretching across a vast region, much in the spirit of the Wildlands Network (www.twp.org). This region spanned from the Mackenzie Mountains of the Canadian Yukon to the Wind River Mountains at the southern end of the Greater Yellowstone Ecosystem in the United States, a monumental distance of some 3,200 kilometers. The inland rainforest was included as a subset of this larger plan. The central part of the inland rainforest was also included in a co-operative project between the USDA Forest Service and the Washington State Department of Transportation, which, in 2003, mapped and modeled core car-nivore habitat and connecting corridors across northern Washington and Idaho and into southern British Columbia.

Although these prior approaches are important to regional conservation, they have many shortcomings, as none of them focus exclusively on inland rainforests. In particular, the rich biological diversity of the inland rainforest, combined with the plight of the mountain caribou, has attracted a growing amount of scientific research funded by universities, government, private foun-dations, and conservation groups for well over a decade now. New studies on lichens, rare plants, mountain caribou, fisheries, grizzly bears, and other large carnivores have given the conservation community a broad range of informa-tion on the entire inland rainforest ecosystem. This has made possible a multi-disciplinary approach to conservation. Thus, in 2003, scientists began identify-ing, optimizing, and proposing candidate protected areas solely within the inland rainforest to identify gaps in protected-area networks. In the United States, conservation-area design (CAD) is currently being used to achieve some of these objectives (see also chapter 2). But ultimately, these are questions re-quiring cross-border coordination and international agreement.

The Building Blocks of CAD

Although the inland rainforest phenomenon is globally rare and unique in North America, it has received little international attention. The overall objec-tive of conservation-area design here, as anywhere else in the temperate and boreal rainforest global network, ought to be based on four well-accepted tenets of conservation: 1) represent ecosystems across their natural range of variation in strictly protected areas; 2) maintain viable populations of native species; 3) sustain ecological and evolutionary processes within an acceptable range of variability; and 4) build a conservation network resilient to environ-mental change, especially climate change (based on Noss and Cooperrider

1994; also see chapter 10). One common conservation-planning approach includes the generally accepted elements of representation, special elements, and focal-species analysis (Noss et al. 2000; see also chapter 2) but primarily emphasizes representation as illustrated above. Another approach concentrates on focal species, assuming that almost all of the critical areas for maintenance of biodiversity, as well as the structure and function of ecological processes, can be identified and protected under the umbrella of an appropriate suite of focal species.

For the CAD, conservation groups have developed computer-based habitat-suitability models to identify core habitat areas needed to sustain a broad suite of focal species, based on their historic fates in the United States. In this case, large carnivores and woodland caribou were selected to represent other rainforest-species needs (Craighead and Cross 2007). Candidate protected areas, capturing large blocks of important habitat and connecting linkages, are then proposed. The resulting proposed reserve network, as applied to the entire inland rainforest region, is interspersed within an otherwise human-dominated landscape that incorporates areas of highest habitat value to the particular focal species and other species into an optimal reserve design (see figure 3-2).

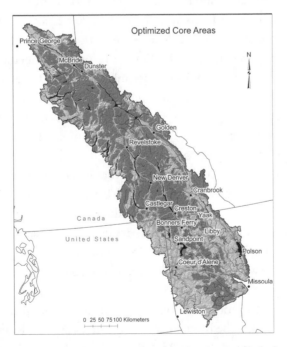

Figure 3-2. Optimized core areas for candidate protected areas (dark shading) in the temperate rainforests of Inland Northwestern North America based on conservation area design (modified from Craighead and Cross 2007).

Based on this approach, scientists incorporated 50–90 percent of the core habitat areas for large carnivores and caribou across the inland rainforest. Although the entire region analyzed included 191,441 square kilometers (19 million hectares), the optimization routine that was used found the most efficient solution through a series of mapping steps. For instance, the proposed terrestrial reserve network of optimized core areas developed by the CAD covered 102,326 square kilometers or ~52 percent of inland rainforests, the minimum that some scientists feel is necessary to save all the pieces. Of this, 17,847 square kilometers or 17.4 percent of the core areas was nonhabitat (primarily rocks and ice). Thus, the actual vegetated land area of habitat cores comprised ~42 percent of the entire inland-rainforest study area (encompassing all vegetation types). A small percentage of the highest-quality aquatic habitat (the best 50 percent) was located outside of the terrestrial cores. These drainages then added an additional 4,584 square kilometers (2.35 percent) to the final CAD solution. With this inclusion, the actual vegetated land area of the optimized habitat cores plus adjoining high-quality watersheds comprised approximately 45 percent of the region.

In developing the CAD, scientists looked at the entire landscape and identified large areas of intact habitat comprised of a mosaic of vegetation and landform types. Further refinements were made by adding taxa to determine if additional area was needed to optimize reserve location; however, this did not change the area required appreciably. Thus, the core areas needing strict protection comprised ~45 percent of the entire inland rainforest (see figure 3-2).

Because ecological processes and species movements often span protected area boundaries, evaluating current reserve networks in relation to surrounding unprotected habitat is important. Connectivity among habitat areas increases the effective size of existing protected areas and plays a critical role in species persistence. In fact, it has long been known that loss of connectivity can lead to localized extinctions (see Lindenmayer and Fischer 2006 for a review). Connectivity of core habitat areas, which include locations well buffered from the influence of human disturbance such as roads and associated development, is important for biotic health and species responses to climate change, with dispersal pathways between suitable habitat areas necessary to ensure species viability over time.

It is important to remember that a CAD is just a blueprint for large-scale conservation planning. The boundaries of core areas and landscape connectivity are both factors that need to be addressed at finer scales with input from local managers and conservationists. For instance, the CAD approach misses north-south connectivity in low elevations. This contrasts directly with land-use plans in some districts of British Columbia (e.g., the Columbia Forest District) that placed all its allotment of old-growth forests for "high-emphasis biodiversity" along the Columbia River Trench. The trench, with its highest point below 800 meters elevation, stretches for hundreds of kilometers between very high, glacier-covered mountain ranges. With changing climates, a low-elevation, north-south trench will likely be a conduit for wildlife to redistribute. The British Columbia land-use allotments, which have been in place due to government planning efforts for more than a decade, are not directly identified by the CAD, since the low-elevation trench is currently highly fragmented by roads and logging. Moreover, some Canadian scientists (personal communication B. McLellan, British Columbia Ministry of Forestry) argue that a CAD approach is not needed in Canada, as large carnivores are abundant and people are more tolerant of them than they historically were in the United States, where carnivores were largely extirpated. Nevertheless, the level of fragmentation throughout the region does call for a more representative network of reserves to reduce competition among deer, moose, and caribou, as well as predator pressure on caribou.

ADDITIONAL THREATS

Like so many of the world's forests, inland rainforests face a multitude of threats from the combination of land use and climate change (see box 3-3). In addition to logging, as already discussed, below we cover some of the key threats to the persistence of this rainforest community.

Hydroelectric Development

A major emerging threat to valley-bottom remnants of inland rainforests are stream-diversion projects for independent power producers. Across both inland and coastal rainforests, especially in British Columbia, hundreds of these "clean energy" projects have begun or been proposed in steep-gradient stream ecosystems, including along waterfall and rapids critical for maintaining core populations of rainforest-dependent organisms like lichens. Diversion of water for hydroelectric production poses a serious threat to riparian ecosystems. These projects are often permitted with minimal environmental review and pose a significant threat to aquatic species that require free-flowing, cool waters (e.g., trout *Oncorhynchus* spp.). It should be noted that road infrastructure associated with the power projects may make it convenient to log some of the otherwise inaccessible portions of the ICH zone, thereby potentially enhancing predator

---

BOX 3-3

Principle Threats Affecting Temperate and Boreal Rainforests of Inland Northwestern North America.

- Approximately 2 million hectares of old-growth forests have been logged since 1960, with logging concentrated in valley bottomlands in productive Interior Cedar-Hemlock rainforest.
- Mountain caribou populations are critically imperiled, with ~1,670 remaining, including only ~34 in the United States, with over 40 percent of the historic range reduced by ecological changes brought on by climate change, wildfire, and especially logging, resulting in excessive caribou mortality from predation and hunting.
- Off-road vehicles and snowmobiles displace wildlife, including caribou.
- Hard-rock mining, hydroelectric dams, and independent hydro-power producers are impacting bull trout (*Salvelinus confluentus*) and other salmonids; licenses can be granted for these and other developments (e.g., ski runs) as exemptions under the government action regulations in British Columbia.
- Fire frequency and intensity is increasing southward, triggered by a warming climate.

---

numbers even further and hence putting even more pressure on mountain caribou.

Climate Change

The likely impacts of climate change in the inland rainforest region include loss of spring and winter snowpack, a reduction and corresponding increase in the number of extremely cold and hot days, respectively, and summer drought and its relationship with fire. This relationship is exemplified by massive fires that occurred during the drought years of the 1890s, 1920s, and 1930s, and again since the late 1980s (Pederson et al. 2009). Such conditions are projected to worsen under a rapidly changing global climate where more frequent droughts and more intense storm activity is anticipated in many places. Recent massive beetle infestations in the dry interior forests of British Columbia that have caused significant tree mortality may be related to warmer winters triggered by global climate change, which is allowing more beetles to survive the Canadian winters (Carroll et al. 2002). In particular, landscape fragmentation from logging and other developments will likely combine with climate-related shifts in drought and fire (southward) that could limit the movement of wildlife species in search of climate refugia.

Significant changes in moisture stress, frequency of fire, snowpack, or growing season could change competitive interactions between rainforest species and ultimately determine which species will dominate and which will vanish. In fact, recent climate trends suggest that a rapid loss of snowpack has occurred over a three-decade period, along with a ~3°C increase in spring temperature (Cayan et al. 2001). For inland rainforests, climate change may also trigger the expansion of drought-adapted species at the expense of moisture-dependent ones in remnant forest patches. The overall sensitivity of inland rainforests to climate change is compounded by the highly fragmented nature of this ecosystem. Steep environmental gradients of moisture and temperature between rainforests and surrounding drier forest types increase fragmentation problems. The long history of road building and logging has considerably expanded the scope of this problem, exacerbating, for example, the invasion of exotic weeds.

CONSERVATION PRIORITIES

Inland rainforests are some of the oldest forest ecosystems in any inland region outside the tropics. Notwithstanding their importance, the overall landscape now bears many scars—as a simple Google Earth virtual flight over the valleys of southern British Columbia makes painfully clear. A significant portion of the oldest rainforest is now gone—most of it logged within the past two decades. It will not grow back—as old growth—under an 80-year timber-harvest rotation or new conditions brought on by climate change. Those forests that remain owe their existence, with few exceptions, to the efforts of a handful of conservation groups, dedicated individuals and, ironically, to the uncertainties regarding mountain caribou, which in turn rely on old forests and the lichens therein.

Notably, greater conservation attention is needed to ensure the continued viability of many "specialist" lichen species currently inhabiting inland rainforests. One point often overlooked is that many rainforest lichens are highly sensitive to atmospheric pollution. Acid rain and industrial pollution have long since caused the decline or loss of many old-growth-dependent lichens once common in eastern North America, western Eurasia and other industrialized regions. By contrast, atmospheric conditions along the Pacific Coast and the inland rainforest region remain for most part very good, hence the forests here continue to support a full "pre-industrial" complement of epiphytic lichens (Goward and Arsenault 2001). In a sense, western North America might be said to have a global responsibility to ensure the continued existence of these sensitive indicators of one of the last vast and nearly pristine forest regions of the world.

The inland rainforests are priceless and irreplaceable archives of the Earth's recent history. Ongoing loss of their oldest stands—antique forests—is a loss not merely to Canadians or Americans, but to the world. Although inland rainforests have been around for thousands of years, their future hinges on a precarious balance between an ecosystem dependent largely on caribou for its protection, disagreements over predator management and what counts as protection, and the uncertainties of climate change. Are inland forests soon to be an artifact of a bygone era when the climate that created them shifts to a drastically different regime? Or is there enough time and political will to fill in the gaps in conservation and to reduce humanity's ever-growing footprint? Indeed, no region should depend largely on a single species to shoulder its conservation burden (see also the northern spotted owl [*Strix occidentalis caurina*] in chapter 2) but rather should address conservation challenges comprehensively, with an eye toward ecosystem management and the future. Inland rainforests are the sum of their ecosystem parts; the most vital of these, old rainforests, mountain caribou, and their connection to rainforest lichens, urgently need stronger protections now.

LITERATURE CITED

Alaback, P. B. 1991. Comparative ecology of temperate rainforests of the Americas along analogous climatic gradients. *Revista Chilena de Historia Natural* 64:399–412.

———, M. Krebs, and P. Rosen. 2000. Ecological characteristics and natural disturbances of interior rainforests of northern Idaho. Pp. 27–37 in *Ecosystem management of forested landscapes.* Ed. Robert G. D'Eon, J. Johnson, and E. Alex Ferguson. Vancouver, BC: Univ. of British Columbia Press.

Apps, C. D., B. N. McLellan, J. G. Woods, and M. F. Proctor. 2004. Estimating grizzly bear distribution and abundance relative to habitat and human influence. *Journal of Wildlife Management* 68:138–52.

Arsenault, A., and T. Goward. 2000. Ecological characteristics of inland rainforests. *Ecoforestry* 15 (4):20–23.

Beschta, R. L., and W. J. Ripple. 2009. Large predators and trophic cascades in terrestrial ecosystems of the western United States. *Biological Conservation.* 142:2401–14.

British Columbia Forest Practices Board. 2004. BC's mountain caribou: Last chance for conservation? Special report. BC Forest Practices Board. FPB/SR/22. Victoria, BC.

Brundtland Commission. 1987. *Our common future: Report of the world commission on environment and development.* Oxford: Oxford Univ. Press.

Carroll, A. L., S. W. Taylor, and J. Regniere. 2002. Effects of climate change on range expansion by the mountain pine beetle in British Columbia. Report prepared for the British Columbia Ministry of Water, Land and Air Protection.

Carstens, B. C., S. J. Brunsfeld, J. R. Demboski, J. M. Good, and J. Sullivan. 2005. Investigating the evolutionary history of the Pacific Northwest mesic forest ecosystem: Hypothesis testing within a comparative phylogeographic framework. *Evolution* 59:1639–52.

Cayan, D. R., S. A. Kammerdiener, M. D. Dettinger, J. M. Caprio, and D. H. Peterson. 2001. Changes in the onset of spring in the western United States. *Bulletin of the American Meteorological Society* 82:399–415.

Craighead, L., and B. Cross. 2007. Identifying core habitat and connectivity for focal species in the interior cedar-hemlock forests of North America to complete a conservation area design. Pp. 1–16 in USDA Forest Service Proceedings RMRS-P-49.

Daubenmire, R. 1943. Vegetational zonation in the Rocky Mountains. *Botanical Review* 9:325–93.

———. 1968. Soil moisture in relation to vegetation distribution in the mountains of northern Idaho. *Ecology* 49:431–38.

———. 1969. Ecological plant geography of the Pacific Northwest. *Madrono* 20:111–28.

Dudley, N., ed. 2008. *Guidelines for applying protected areas management categories.* Gland, Switzerland: IUCN.

Edwards, R. Y. 1954. Fire and the decline of a mountain caribou herd. *Journal of Wildlife Management* 18:521–26

———, J. Soos, and R. W. Ritcey. 1960. Quantitative observations on epidendric lichens used as food by caribou. *Ecology* 41:425–31.

Gavin, D. G., F. S. Hu, I. R. Walker, and K. Westover. 2009. The Northern Inland temperate rainforest of British Columbia: Old forests with a young history? *Northwest Science* 83:70–78.

Goward, T. 1994. Notes on old-growth-dependent epiphytic macrolichens in the humid old-growth forests in inland British Columbia, Canada. *Acta Botanica Fennica* 150:31–38.

———. 1998. Observations on the ecology of the lichen genus Bryoria in high elevation conifer forests. *Canadian Field Naturalist*: 112:496–501.

———, and J. Pojar. 1998. Antique forests and epiphytic macrolichens in the Kispiox Valley. *Forest Sciences Extension Note 33.* British Columbia Ministry of Forests, Smithers, BC.

———, and A. Arsenault. 2000. Inland old-growth rainforests: Safe havens for rare lichens? Pp. 759–766 in *Proceedings of a conference on the biology and management of species and habitats at risk, Kamloops, B.C., 15–19 February, 1999.* Ed. L. Darling. B.C. Ministry of Environment, Lands and Parks, Victoria, BC, and University College of the Cariboo, Kamloops, BC.

———, and A. Arsenault. 2001. Cyanolichens and conifers: Implications for global conservation. *Forest, Snow and Landscape Research* 75:303–18.

———. 2003. On the vertical zonation of hair lichens (Bryoria) in the canopies of high-elevation old-growth conifer forests. *The Canadian Field-Naturalist* 114:39–43.

————, and T. Spribille. 2005. Lichenological evidence for the recognition of inland rain forests in western North America. *Journal of Biogeography* 32:1209–19.

————, and C. Björk. 2009. Wilfred Schofield: A waterfall tribute. *Botanical Electronic News* 404. ISSN 1188-603X.

Hackman, A., and P. Paquet. 1994. Large carnivore conservation in the Rocky Mountains: A long-term strategy for maintaining free-ranging and self-sustaining populations of carnivores. World Wildlife Fund, Canada. Toronto.

Hebblewhite, M., C. White, C. Nietvelt, J. Mckenzie, T. Hurd, J. Fryxell, S. Bayley, and P. C. Paquet. 2005. Human activity mediates a trophic cascade caused by wolves. *Ecology* 86:1320–30.

Hebda, R. J., and R. W. Mathewes. 1984. Holocene history of cedar and native Indian cultures of the North American Pacific Coast. *Science* 225:711–13.

————. 1995. British Columbia vegetation and climate history with focus on 6 ka bp. *Géographie Physique et Quaternaire* 49:55–79.

IUCN (World Conservation Union). 2008. Guidelines for applying protected area management categories: draft of revised guidelines. Gland, Switzerland.

Johnson, F. D., and R. Steele. 1978. New plant records for Idaho from Pacific coastal refugia. *Northwest Science* 52:205–11.

Kinley, T. A., T. Goward, B. N. McLellan, and R. Serrouya. 2007. The influence of variable snowpacks on habitat use by mountain caribou. *Rangifer*, Special Issue 17:93–102.

Lawford, R., P. Alaback, and E. R. Fuentes, eds. 1996. *High-latitude rain forests and associated ecosystems of the west coast of the Americas: Climate, hydrology, ecology and conservation.* Ecological Studies 116. Berlin: Springer-Verlag.

Lindenmayer, D. B., and J. Fischer. 2006. *Habitat fragmentation and landscape change: An ecological and conservation synthesis.* Washington, DC: Island Press.

Lorain, C. C. 1988. Floristic history and distribution of coastal disjunct plants of the northern Rocky Mountains. M.S. Thesis, University of Idaho, Moscow, ID.

McKenney, D. W., J. H. Pedlar, K. Lawrence, K. Campbell, and M. F. Hutchinson. 2007. Potential impacts of climate change on the distribution of North American Trees. *BioScience* 57:939–48.

Ministry of Forests and Range. 2008. Biogeoclimatic zones of British Columbia. Vancouver, BC. www.for.gov.bc.ca/hfd/pubs/Docs/M/M008.htm

Moola, F., D. Page, M. Connolly, and L. Coulter. 2007. Waiting for the ark: The biodiversity crisis in British Columbia, Canada, and the need for a strong endangered species law. *Biodiversity* 8 (1):3–11.

Mountain Caribou Science Team. 2005. Mountain caribou in British Columbia: A situational analysis. British Columbia Ministry of Environment, Species at Risk Coordination Office. www.env.gov.bc.ca/sarco/

Mountain Caribou Technical Advisory Committee. 2002. A strategy for the recovery of mountain caribou in British Columbia. British Columbia Ministry of Water, Land, and Air Protection. Victoria, British Columbia.

Mowat, G., and C. Strobeck. 2000. Estimating population size of grizzly bears using hair capture, DNA profiling, and mark-recapture analysis. *Journal of Wildlife Management* 64 (1):183–93.

————, D. C. Heard, D. R. Seip, K. G. Poole, G. Stenhouse, and D. Paetkau. 2005. Grizzly *Ursus arctos* and black bear *U. americanus* densities in the interior mountains of North America. *Wildlife Biology* 11:31–48.

Newmaster, S. G., R. J. Belland, A. Arsenault, and D. H. Vitt. 2003. Patterning of bryophyte diversity in humid coastal and inland cedar-hemlock forests of British Columbia. *Environmental Review* 11:159–85.

Noss, R. F., and A. Y. Cooperrider. 1994. *Saving nature's legacy.* Washington, DC: Island Press.

————, J. R. Strittholt, G. E. Heilman Jr., P. A. Frost, and M. Sorensen. 2000. Conservation planning in the redwoods region. Pp. 201–28 in *The redwood forest.* Ed. R. F. Noss. Washington, DC: Island Press.

Orians, G. H., P. A. Cochran, J. W. Duffield, T. K. Fuller, R. J. Gutierrez, W. M. Hanemann, F. C. James, et al. 1997. *Wolves, bears, and their prey in Alaska. Biological and Social Challenges in Wildlife Management,* Committee on Management of Wolf and Bear Populations, in Alaska Board on Biology Commission on Life Sciences, National Research Council. Washington, DC: National Academy Press.

Paquet, P. C., Alexander, S., Donelon, S., and C. Callaghan. Forthcoming. Influence of anthropogenically modified snow conditions on movements and predatory behaviour of gray wolves. In *The world of wolves: New perspectives on ecology, behaviour, and policy.* Ed. Musiani, M., L. Boitani, and P. Paquet. Calgary, AB: Univ. of Calgary Press.

Pederson, G. T., L. J. Graumlich, D. B. Fagre, T. Kipfer, and C. C. Muhlfeld. 2009. A century of climate and ecosystem change in western Montana: What do temperate trends portend? Climate Change. Published online, August 21, 2009. Springer Science.

Radies, D. N., and D. S. Coxson. 2004. Macrolichen colonization on 120–140 year old *Tsuga heterophylla* in wet temperate rainforests of central-interior British Columbia: A comparison of lichen response to even-aged versus old-growth stand structure. *Lichenologist* 36:235–47.

Rominger, E. M., C. T. Robbins, and M. A. Evans. 1996. Winter foraging ecology of woodland caribou in northeastern Washington. *Journal of Wildlife Management* 60(4):719–28.

Sanborn, P., M. Geertsema, A. J. T. Jull, and B. Hawkes. 2006. Soil and sedimentary charcoal evidence for Holocene forest fires in an inland temperate rainforest, east-central British Columbia, Canada. *The Holocene* 16:415–27.

Scott, J. M., F. Davis, B. Csuti, R. F. Noss, B. Butterfield, C. Groves, H. Anderson, et al. 1993. Gap analysis: A geographic approach to protection of biological diversity. *Wildlife Monographs* 123.

Smith, J., and W. C. Fischer. 1997. Fire ecology of the forest habitat types of Northern Idaho. General Technical Report INT-GTR-363, US Department of Agriculture (USDA) Forest Service, Intermountain Research Station, Ogden, UT.

Spalding, D. J. 2000. The early history of woodland caribou (*Rangifer tarandus caribou*) in British Columbia. *Wildlife Bulletin* B-100. Ministry of Environment, Lands and Parks. Victoria, BC.

Spribille, T. 2002. The mountain forests of British Columbia and the American Northwest: Floristic patterns and syntaxonomy. *Folia Geobotanica* 37:475–508.

————, C. R. Björk, S. Ekman, J. A. Elix, T. Goward, C. Printzen, T. Tønsberg, and T. Wheeler. 2009. Contributions to an epiphytic lichen flora of northwest North

America. I. Eight new species from British Columbia inland rainforests. *The Bryologist* 112 (1):109–37.

Stronen, A., P. C. Paquet, S. Herrero, S. Sharpe, and N. Waters. 2007. Translocation and recovery efforts for the Telkwa caribou, *Rangifer tarandus caribou*, herd in westcentral British Columbia, 1997–2005. *Canadian Field-Naturalist* 121:155–63.

Swetnam, T. W., and A. M. Lynch. 1993. Multi-century, regional-scale patterns of western spruce budworm history. *Ecological Monographs* 63:399–424.

Terborgh, J., J. A. Estes, P. C. Paquet, K. Ralls, D. Boyd-Heger, B. J. Miller, and R. F. Noss. 1999. The role of top carnivores in regulating terrestrial ecosystems. Pp. 39–64 in *Continental Conservation*. Ed. M. E. Soule and J. Terborgh. Washington, DC: Island Press.

Wilson, S. F. 2009. Recommendations for predator-prey management to benefit the recovery of mountain caribou in British Columbia. Report prepared for BC Ministry of Environment, Victoria, BC.

Wittmer, H. U., B. N. McLellan, D. R. Seip, J. A. Young, T. A. Kinley, G. S. Watts, and D. Hamilton. 2005. Population dynamics of the endangered mountain ecotype of woodland caribou (*Rangifer tarandus caribou*) in British Columbia, Canada. *Canadian Journal of Zoology* 83:407–18.

CHAPTER 4

Perhumid Boreal and Hemiboreal Forests of Eastern Canada

Stephen R. Clayden, Robert P. Cameron, and John W. McCarthy

Eastern Canada takes in the rugged easternmost prominence of North America adjoining the North Atlantic Ocean. From south to north, and along gradients of elevation and distance from the ocean, it encompasses a range of temperate to boreal forests. Distinctive variants of these forests occur in the wettest coastal and montane areas of the region. They occur discontinuously from the Atlantic Coast of Nova Scotia to southern Labrador (43–52°N latitude), and from the Avalon Peninsula on the island of Newfoundland to the Appalachian and Laurentian highlands of eastern Quebec (53–73°W longitude).

In this chapter, we briefly describe the climatic context, composition, dynamics, and conservation status of these forest communities. Although research is needed to refine their definition and survey their variation along the regional temperate to boreal bioclimatic gradient, we propose that they deserve consideration as rainforests. We are not the first to do so. Holien and Tønsberg (1996), for example, suggested that boreal forests dominated by balsam fir (*Abies balsamea*) in the wettest, least fire-susceptible areas of Eastern Canada can be appropriately categorized as boreal rainforests. Thompson et al. (2003) demonstrated that old-growth stages in particular of these fir stands are distinct in structure and composition; they termed these communities "wet boreal forests," characterizing them as "conifer-dominated forests that receive sufficient moisture from precipitation and fog, especially during summer, such that fires are rare to nonexistent." However, the term "wet boreal" does not adequately differentiate ordinary wet forests occurring in poorly drained situations from those restricted to areas of high precipitation.

PERHUMIDITY AND RAINFORESTS

Here, we use the term "perhumid" to designate the wettest climates and their associated forests in Eastern Canada. We use it in the specific sense of Thornthwaite (1948), who first defined a perhumid climate as one in which precipitation much exceeds evaporation and plant transpiration, resulting in year-round wetness. Perhumid areas may undergo short periods during an average year when water losses exceed water gains. However, abundant rain or snow and snowmelt during the flanking periods make up for these shortfalls. In classifying climate types according to the balance between precipitation and evapotranspiration, Thornthwaite (1931) drew attention to their correlation with major vegetation and soil formations. He identified rainforest as the characteristic expression of climates with the largest and most continuous water surpluses. Such surpluses occur not only in the tropics, but also along poleward and elevational thermal gradients.

Eastern Canada and the adjoining northeastern United States take in one of the largest areas of plentiful, seasonally equable precipitation occurring anywhere in the world (Hare 1961). With evapotranspiration rates much lower than those in the tropics, this region also has extensive areas meeting the threshold value for perhumidity as defined in the moisture index of Thornthwaite (Feddema 2005). When this index is re-scaled to provide more differentiation of the wettest climates, areas such as the Pacific Coast of North America and coastal Norway stand out as having exceptionally high values (Feddema 2005).

In eastern North America, perhumid climates occur in island-like areas of high elevation along the Appalachian Mountain range, and more widely in coastal Eastern Canada (Thornthwaite 1948; Sanderson 1948; Phillips et al. 1990; see figure 4-1). Shanks (1954) demonstrated that high-elevation spruce-fir stands in the Great Smoky Mountains of the southern Appalachians occupy a highly perhumid "rainforest climate" as defined by Thornthwaite (1931). These communities continue to be categorized as rainforests or cloud forests by some ecologists (e.g., Reinhardt and Smith 2008). Shanks (1954) also observed that the closest low-elevation analogue of the climate and vegetation of the spruce-fir zone in the Smoky Mountains is in coastal eastern Maine (USA) and adjacent New Brunswick (Canada), about 1,600 kilometers to the northeast.

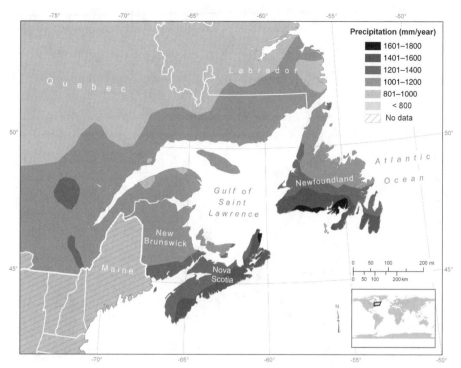

Figure 4-1. Mean annual precipitation in Eastern Canada, based on normals for 1971–2000 (Environment Canada 2002). Small montane areas with totals greater than 1,400 millimeters per year are too small to be show at the map scale.

The rainforest climate model presented in Chapter 1 is based on a multivariate analysis of climate data from areas previously deemed, on botanical and other evidence, to support temperate or boreal rainforests. It did not include data from Eastern Canada. However, the model predicts extensive areas of potential rainforest occurrence in this region. Those areas total about 6 million hectares, a value quite similar to that obtained by overlaying the perhumid areas on the Thornthwaite map (see figure 4-2) and regional forest cover data (see chapter 1). However, there are considerable differences between the locations and outlines of the modeled areas and the actual occurrence of forests that seem to us to approach most closely to what has been categorized as rainforest in other mid- to high-latitude regions. This said, we also consider that the map of moisture regions provides only a rough approximation, though a better one, of the potential occurrence of such forests in Eastern Canada. This map is likely to need the most revision in areas where there are few long-term climate stations. We also emphasize that there are other climatic variables, in addition to moisture, that influence

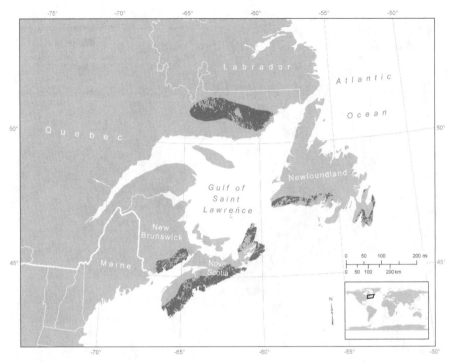

Figure 4-2. Moisture regions in Eastern Canada, adapted from Phillips et al. (1990). Only the wettest category of the Thornthwaite (1948) moisture index for perhumid (index values >100) is shown and was superimposed onto the Canadian vegetation- and land-cover data set.

the composition and distribution of the most rainforest-like communities in Eastern Canada (see below).

INSIDE THE REGIONAL CLIMATE

Annual precipitation in Eastern Canada is mostly in the range of 1,000 to 1,600 millimeters, though locally higher in some coastal and montane areas (see figure 4-1). The general wetness of Eastern Canada is a consequence of its location in relation to major airstreams and ocean currents. The westerly, continental airflows that dominate the region draw in arctic and subtropical airstreams on their north and south flanks (Hare and Thomas 1974). Fronts form as the moist subtropical flows override cooler northern air, generating frequent precipitation. Strong winds and abrupt temperature changes are also common.

Coastal areas that intercept humid northeast-tracking air masses have the highest precipitation and moisture-index values in the region. These include the north shore of the Bay of Fundy, the Atlantic slope of Nova Scotia, and southern Newfoundland (see figure 4-1). Upwards of 85 percent of the precipitation in these areas falls as rain (Cameron et al. 2009). Inland, this proportion decreases to 65 to 75 percent. At the highest elevations, snow accounts for about half of all precipitation. Measureable precipitation (greater than or equal to 0.2 millimeters), at least as recorded at climate stations, occurs less frequently than in the boreal rainforest climates of northwestern Europe (see below).

Fog is an important contributor to the water budgets of the cooler coastal areas. Along the Fundy and Atlantic coasts of New Brunswick and Nova Scotia, fog water deposition increases recorded precipitation by an estimated 5 to 8 percent, while evapotranspiration is correspondingly reduced (Yin and Arp 1994). Southern and eastern Newfoundland have one of the foggiest sea-level climates in the world, resulting from the convergence of the cold Labrador Current and the warm Gulf Stream over the adjoining Grand Banks.

Orographic lifting and cooling of air masses increase precipitation over high ground in the region, especially in the Eastern Townships, Laurentian Highlands, and Chic-Choc Mountains of Quebec, the Kedgwick and Miramichi Highlands of New Brunswick, the Cape Breton Highlands of Nova Scotia, and the Long Range Mountains of western Newfoundland. Elevations in these areas range from about 500 meters to a maximum of 1,268 meters in the Chic-Chocs. The higher elevations are often immersed in the cloud base, increasing moisture interception and reducing evapotranspiration in a manner analogous to that occurring in foggy coastal areas. Clouds contain more water than sea-level fog (Yin and Arp 1994). Cloud water interception by montane spruce-fir communities in the Appalachians accounts for 20 to 50 percent of their water inputs (Mohnen 1992). Comparable inputs likely occur in some montane areas of Eastern Canada (Schemenauer et al. 1995), though these have not been quantified.

The perhumidity of the cold plateau north of the Gulf of St. Lawrence (see figure 4-2) apparently results from the combined influences of elevation, proximity to the ocean, and orientation toward the track of moist weather systems. This area has no inland climate stations for which long-term normals are available. However, high precipitation can be inferred from the heavy runoff in the major rivers draining the plateau (Hare and Thomas 1974).

For its latitude, Eastern Canada has an unusually cool climate (Tuhkanen 1984; Clayden 2010). This is due in part to the dominance of continental over-oceanic airstreams, and in part to the refrigerating influence of the Labrador Current, which transports arctic water around the outer coasts of the region. Mean annual temperatures at low elevation range from a maximum of 7.5°C in southern Nova Scotia to 0.5°C near the northern tip of the island of New-foundland (Environment Canada 2002). The plateau of the Chic-Choc Mountains in the Gaspé Peninsula of Quebec has means of −3 to −5°C, cold enough to sustain a disjunct body of permafrost (Gray and Brown 1979). North of the Gulf of St. Lawrence, mean annual temperature is about 1°C at the coast, falling to −3°C or colder on the inland plateau.

In a zonal bioclimatic context, Eastern Canada is centered on the eastern-most portion in North America of the transition between temperate and boreal zones. The warmest areas of interior southern New Brunswick and Nova Scotia are "north temperate" in the zonal system developed by Fennoscandian bio-geographers (e.g., Ahti 1964; Hämet-Ahti 1980; Tuhkanen 1984). The boreal zone in this system is divided into southern, middle and northern subzones, with hemiarctic and hemiboreal transitional zones to the north and south. Most of Eastern Canada falls into the hemiboreal and southern boreal zones, but the colder parts of Newfoundland and areas north of the Gulf of St. Lawrence are middle to northern boreal (Damman 1983; Clayden 2010).

Because the dominant airflows over the region are continental, not oceanic, seasonal temperature variation is greater than in the coastal rainforest climates of western North America and western Europe. Regional temperatures are moderated, nonetheless, by proximity to the Atlantic Ocean and Bay of Fundy. These remain unfrozen in winter, but cooled in summer by tidal mixing and the influence of the Labrador Current. Sea temperatures are more stratified in the Gulf of St. Lawrence, and it thus develops extensive ice cover in winter.

The difference between the mean temperatures of the warmest and coldest months of the year is a common measure of the continentality or oceanicity of a climate. It is one of the key parameters correlated with rainforest occurrence in northwestern Europe (Holien and Tønsberg 1996). Based on the "Conrad index," which adjusts the annual temperature range for differences in the latitudes of areas being compared, the most thermally oceanic parts of Eastern Canada are outside the envelope of values associated with coastal rainforests in western Europe and western North America.

Tuhkanen (1984) divided the overall range of Conrad index values in the boreal and adjoining bioclimatic zones into seven sectors, ranging from extremely continental (C3) to extremely oceanic (O3). Most of Eastern Canada is in the intermediate OC sector. Most of the island of Newfoundland, much of Nova Scotia's south coast and Cape Breton Island, and the coastal fringe along the Bay of Fundy in New Brunswick are in the somewhat oceanic O1 sector (Clayden 2010). Inland montane areas in the region have more continental (OC) thermal conditions, though they also have high precipitation totals (locally above 1,400 millimeters in a number of areas). The coastal rainforests of western North America and Europe are in the highly oceanic O2 and O3 sectors, which have no representation in Eastern Canada. However, the inland rainforests of British Columbia, and marginal areas of boreal rainforest in Norway, are in the O1 sector (Goward and Ahti 1992; Holien and Tønsberg 1996; Goward and Spribille 2005).

TEMPERATE TO BOREAL GRADIENT IN EASTERN CANADIAN PERHUMID FORESTS

Some important features of perhumid coastal and montane forests in Eastern Canada are summarized in Table 4-1. The northernmost coastal stands are boreal communities dominated by balsam fir, often with a component of black spruce (*Picea mariana*), white spruce (*P. glauca*), or heartleaf birch (*Betula cordifolia*, plate 5). These occur in areas bordering the Atlantic Ocean and Gulf of St. Lawrence (Thompson et al. 2003). They are most extensively developed on the island of Newfoundland (Damman 1983), with transitional boreal-hemiboreal variants along the eastern Atlantic shore of Nova Scotia (Loucks 1962; Clayden 2010). The wettest montane, fir-dominated forests in the Cape Breton Highlands (Loucks 1962), Eastern Townships (Marcotte and Grandtner 1974), Laurentian Highlands (Desponts et al. 2002) and Gaspé Peninsula (Desponts et al. 2004) are similar in structure and dynamics to the low-elevation

Table 4-1. Some globally significant attributes of perhumid coastal and montane forests of temperate-to-boreal Eastern Canada.

Attribute	Importance
Large area of abundant, seasonally equable precipitation; frequent coastal fog	Diverse perhumid forests; extensive carbon-sequestering peatlands; numerous high-volume rivers with important Atlantic salmon (*Salmo salar*) populations
Oldest known individuals of balsam fir and red spruce	Reservoir of genetic diversity
Rich lichen biota, including many rare species and nearly all of the global population of the endangered boreal felt lichen	Sensitive indicators of environmental health and change; protection of rare species that are more highly threatened in other regions
Northern end of corridor of perhumid climates and vegetation linking tropical- and temperate-montane with northern coastal biota	High diversity of migrant neotropical songbirds; northernmost populations of other biota with tropical-montane affinities

coastal stands. However, with the exception of the Cape Breton Highlands, their thermal conditions are more continental than those of the coastal forests. They closely resemble the montane balsam fir forests of the northeastern United States (Reiners and Lang 1979).

With increasing age, the fir-dominated forests in the most oceanic (O1) coastal areas of the region acquire a high diversity and biomass of lichens on living and dead trees, provided that desiccating winds are not too strong. Their understories are dominated by bryophytes, often including *Rhytidiadelphus loreus* (lanky moss); this species is absent from balsam fir forests in more continental areas. Epiphytic bryophytes have low diversity and biomass in these stands, in contrast to their abundance in rainforests of the Pacific Coast of North America. Interestingly, one of the common epiphytic mosses of west-coast rainforests, *Antitrichia curtipendula* (hanging wing moss), has its only known eastern North American occurrences in the most oceanic and foggy area of southern Newfoundland. Here, however, it occurs on the ground in rocky moss- and lichen-dominated heath, where the only trees are patches of dwarfed, wind-sheared fir krummholz, in Newfoundland known as tuckamore.

Owing to their wetness, the coastal and montane fir-dominated forests have either no history of forest fire or much longer fire cycles than boreal forests in more continental regions farther west (Wein and Moore 1979; Foster 1983; McCarthy 2001; Lauzon et al. 2007; Bouchard et al. 2008). Insect epidemics, fungal diseases, and wind are the main factors regulating stand dynamics. Average gap size and stand age-class structure vary, depending on the scale and frequency of such disturbances. Sporadic small disturbances are prevalent (McCarthy 2001), yielding continuous multi-aged stands at the scale of the landscape. Although small in stature, with mature fir trees rarely exceeding 15 meters in height, these forests acquire and maintain many old-growth features (Thompson et al. 2003; McCarthy and Weetman 2006). They have been considerably reduced in extent and age by clear-cut logging, which creates gaps much larger than those resulting from most natural disturbances (Thompson et al. 2003).

Wind damage occasionally occurs on a much larger scale in Eastern Canada. The most powerful winds are those associated with tropical storms and hurricanes that hit the region in late summer and fall. The most impressive of these systems in recent years was Hurricane Juan in 2003, which affected forested areas over nearly one-third of Nova Scotia. In November 1994, hurricane-force winds leveled an estimated 17,000 hectares of mature balsam fir forest in the boreal highlands of north-central New Brunswick (Zelazny 2007). It is characteristic of the regeneration of these communities, as of the coastal fir forests, that the early stages of stand development are most often soon domi-nated by a high density of young balsam fir trees. Prior to stand-replacing or gap-forming disturbances, these remained suppressed in the forest understory. They later contribute to the high volume of snags and coarse woody debris characteristic of old fir stands.

In the transition between boreal and temperate bioclimatic zones in East-ern Canada, a range of deciduous and coniferous trees assume dominance. In forest classifications, this transition zone is known as the Acadian Forest Region (Rowe 1972). In the perhumid portion of the regional moisture gradient, mixed or coniferous forests dominated by red spruce (*P. rubens*) are most fre-quent on mesic sites. The fire cycle in these stands near the Bay of Fundy is over 1,000 years (Wein and Moore 1977). They have varying admixtures of balsam fir, yellow birch (*B. alleghaniensis*), heartleaf birch, red maple (*Acer rubrum*), and mountain ash (*Sorbus americana*). Mountain wood fern (*Dryopteris campyloptera*), wood sorrel (*Oxalis montana*), and whorled wood aster (*Oclemena acuminata*) are characteristic understory species, though continuous thick carpets of *Bazzania trilobata* (three-lobed bazzania) and other bryophytes are formed in the oldest, most humid stands. These forests are very similar in composition to red spruce–dominated stands occurring at high elevation in the northern Appalachians (White and Cogbill 1992).

Perhumid red spruce forests are well developed along the Fundy Coast, particularly in deep ravines protected from the strongest winds—protected, too, from the intensive logging that has reduced old-growth forest on more accessi-ble terrain to a tiny fraction of its former extent in the Acadian Forest Region (Mosseler et al. 2003). The oldest red spruce known anywhere in the range of this species, a tree about 450 years old, is in such a stand in New Brunswick (Phillips and Laroque 2006). Where old yellow birches are present in the mixed coastal forests, their bryophyte- and lichen-covered trunks are sometimes also colonized by a few vascular plant species otherwise confined to the forest floor or to shaded rocky outcrops. The most frequent of these epiphytes are polypody ferns (*Polypodium appalachianum–P. virginianum* complex).

The elevation-related and inland-coastal gradients of vegetation in the wet climatic sectors of the region are broadly similar to one another (Clayden 2010). These gradients also mirror the topographic gradient of climate and veg-etation occurring in the Appalachian Mountains in the eastern United States (reviewed by Cogbill and White 1991). Fir stands dominate the highest eleva-tions and coolest coastal areas; below these, and on somewhat warmer coasts, red spruce is dominant, often with yellow birch. Next in sequence are de-ciduous or mixed forests with more distinctly temperate trees including beech

(*Fagus grandifolia*), sugar maple (*A. saccharum*), and eastern hemlock (*Tsuga canadensis*). In New Brunswick and Nova Scotia, the latter trees enter into the margins of areas with perhumid climates. Luxuriant, species-rich communities of epiphytic bryophytes and lichens are often present on hardwood trees in old-growth remnants of these mixed or deciduous forests (see plate 5b).

Whether the warmest (north temperate) bioclimatic zone represented in the region has a distinctly perhumid sector and associated forests is not yet clear. The main candidate area is the interior of southern Nova Scotia. Liverpool Big Falls may be a representative climate station (Environment Canada 2002), with a July mean temperature of 19.5°C and mean annual precipitation of 1,517 millimeters. July rainfall averages 100 millimeters. The annual range of monthly mean temperatures places this area in the intermediate sector (OC) along the oceanic-continental gradient. In spite of its high precipitation, the area is characterized by thin, coarse-textured soils. It is regularly affected by hurricane winds, and has a long fire history (Loucks 1962). Thus, old forests are uncommon.

COMPARISON WITH WESTERN NORTH AMERICAN AND WESTERN EUROPEAN RAINFORESTS

The contrasting stature and tree species diversity of western European and western North American rainforests are, in large part, artifacts of their contrasting histories. A number of now-extinct temperate trees (including, for example, species of *Pseudotsuga*, *Tsuga*, *Sequoia*, and *Thuja*) were present in Europe until the mid- or late-Tertiary period. They disappeared as a consequence of the long-term cooling that culminated in the Pleistocene glaciations (Dahl 1998). These changes caused vegetation zones and species ranges to be displaced southward throughout the Northern Hemisphere. In North America, the western and eastern (Appalachian) cordilleras provided north-south corridors along which these shifts could take place. In Europe, the east-west orientation of the major mountain ranges posed a barrier to southward migration of the biota, resulting in higher rates of extinction. The modern absence in Europe of trees specially adapted to rainforest climates, in contrast to their persistence in the Pacific Northwest, enabled the occupation of rainforest niches by generalist, wide-ranging species such as Norway spruce (*Picea abies*) and Scots pine (*Pinus sylvestris*).

Eastern North America has a greater diversity of tree species than either Europe or western North America. Like Europe, however, it lacks rainforest-specialist trees, attesting to the absence, over a long evolutionary time frame, of climates as persistently moderate and wet as those of the Pacific Coast (see also chapter 1). There, a rainforest formation of variably tropical, subtropical, temperate, and boreal affinities has occurred more or less continuously since the early Tertiary period (Graham 1999). The shorter average life spans of eastern versus Pacific North American trees similarly imply long-term adaptation to shorter average intervals of destructive (i.e., life-ending) disturbances, probably driven largely by climatic differences.

The moist climates of the Appalachian Mountains and coastal regions of eastern North America are nonetheless an ancient and continuing evolutionary arena. During the Pleistocene glaciations, the southern Appalachians provided a refugium within which several tree species diverged from their more wide-ranging boreal ancestors. These include red spruce, a descendant of eastern populations of black spruce (Perron et al. 2000), varieties of balsam fir with conspicuous cone-bracts (*A. balsamea* var. *fraseri* and var. *phanerolepis*) (Potter et al. 2010), and heartleaf birch. These trees are distinctive components of perhumid montane and coastal forests in the Appalachians and Eastern Canada (Clayden 2000), with more-sporadic occurrences in areas of high precipitation as far west as the Great Lakes. Fraser fir (*A. balsamea* var. *fraseri*) is confined to the southern Appalachians.

LICHENS AS INDICATORS OF RAINFOREST CONDITIONS

The most striking indicators of ecological parallelism between the perhumid and oceanic forests of Eastern Canada and those that have been categorized as rainforests (particularly montane rainforests or cloud forests) in other regions, are their epiphytic lichens. A description of Ecuadorian cloud forest or "upper montane rainforest" as "exceptionally rich in lichens, especially species of *Erioderma*, *Everniastrum*, *Heterodermia*, *Hypotrachyna*, *Leptogium*, *Sticta*, and *Usnea*" (Arvidsson 1991), could well apply to some coastal, old-growth, lichen-rich forests in Nova Scotia and New Brunswick. Other lichen genera characteristic of Central and South American cloud forests include, for example, *Coccocarpia*, *Lobaria*, *Menegazzia*, *Normandina*, *Pannaria*, *Parmeliella*, *Parmelinopsis*, *Parmotrema*, and *Pseudocyphellaria* (Arvidsson 1991).

All of these genera are represented in Eastern Canada by species shared with Appalachian and tropical montane forests. In a number of cases, their low-elevation, coastal northeastern North American populations are widely disjunct from the next nearest populations in high-elevation spruce-fir or mixed forests in the Appalachians (Gowan and Brodo 1988; Clayden 2010).

Disjunct occurrences of several montane and oceanic lichens are also known from humid boreal and hemiboreal forests near Lake Superior in the North American interior. Holien and Tønsberg (1996) accordingly suggested that these stands, too, could be categorized as rainforests.

The shared presence of this assemblage of lichens in widely separated areas implies certain commonalities in the environments supporting them. It is also indicative of the impressive capacity of many lichens for long-distance dispersal, enabling them to colonize isolated areas of suitable habitat. It is not only a general regime of temperature and precipitation conditions that is a requirement for their occurrence in Eastern Canada. The smaller-scale structural, dynamic, and microclimatic characteristics of the old, wet forests that provide their local niches are also key requirements (see also chapter 3 for a similar perspective). These niches are evidently replicated across tropical- or temperate-montane and northern-oceanic forests that otherwise differ greatly in floristic composition. The thermal differences in their climates are profound, the one region with little or no seasonality, the other with sharply defined winters and summers. But without a habitat that in some respects is rainforest-like or cloud forest–like, a number of these lichens would be absent, or largely so, from Eastern Canada.

Holien and Tønsberg (1996) observed that both tropical and extra-tropical rainforests can be characterized by the dependence of their biota, from the soil to the tree canopy, on constant high humidity. It was on this basis that they proposed the recognition of lichen-rich spruce-dominated forests in wet, highly oceanic areas of Norway as boreal rainforests (see chapter 6). Most distinctive among the lichens of these forests is a group of about 15 northern species that have their only or their main European occurrences in the Trøndelag region of central Norway. About half of the lichens in this "Trøndelag element" also occur in Eastern Canada and other oceanic regions of the northern hemisphere: *Bryoria trichodes* spp. *americana*, *Cavernularia hultenii*, *Erioderma pedicellatum*, *Fuscopannaria ahlneri*, *Gyalideopsis piceicola*, *Lecidea roseotincta*, and *Platismatia norvegica*.

Wet fir-dominated forests on the island of Newfoundland have the richest representation of these boreal oceanic species (Ahti 1983). Most of them are also present in wet coniferous forests along the coolest portions of the Atlantic shore in Nova Scotia. On the other hand, several of the predominantly temperate and tropical-montane lichen genera noted above are absent from Newfoundland, which has at most a marginal representation of hemiboreal climates and vegetation (Ahti 1983; Damman 1983; Tuhkanen 1984; Clayden 2010). Some of the Trøndelag species, including *Platismatia norvegica* and *Gyalideopsis piceicola*, extend from oceanic coastal into the most highly perhumid montane fir forests in Eastern Canada; *G. piceicola* and *L. roseotincta* have been discovered recently in high-elevation spruce-fir forests in the southern Appalachians (Tønsberg 2006). The globally rare boreal felt lichen, *Erioderma pedicellatum* (see box 4-1), is a boreal representative of a primarily tropical genus. Its range, like those of many other oceanic lichens in Eastern Canada, is suggestive of the importance of the Appalachian cordillera as a long-standing, perhumid, migrational corridor linking tropical-montane and high-latitude coastal environments.

BOX 4-1

Ecology of Boreal Felt Lichen in Eastern Canada.

Boreal felt lichen is a globally endangered lichen known from only a few northern coastal areas of the world. It inhabits boreal and hemiboreal rain-forests, with Eastern Canada harbouring most of its known population (Maass and Yetman 2002). It is a cyanolichen, comprising a fungus and a cyanobacterium. These symbiotic partners form a grey, leafy thallus, several or more centimeters in width, which turns dark green when wet. It grows most often on balsam fir, occasionally on black spruce or red maple. It is never found far from wetlands, and usually holds close to the rainy, fog-bound coast.

Boreal felt lichen is threatened by air pollution and commercial forestry, and its populations continue to decline in Eastern Canada. It is among the lichens most sensitive to human disturbances. It thus serves as an early warning of human impacts on rainforest ecosystems.

Newfoundland has the largest population of boreal felt lichen: at least 10,000 individuals. However, recent modeling indicates, unfortunately, that the population is declining. Researchers suspect that acid rain may be a contributing factor as well as loss of habitat from tree browsing by introduced moose.

Only 180 thalli are known in Nova Scotia. Although new occurrences are being found by researchers, old ones are disappearing (Cameron and Neily 2008). Boreal felt lichen is believed to be extirpated from New Brunswick as well as from Sweden, where logging eliminated the last known population (single thallus). Acid rain and fog (Cox et al. 1996) have possibly degraded the habitat to a point where it cannot survive in New Brunswick any longer. Recent finds in Alaska may hold promise for the future of this rare lichen.

LEVELS OF PROTECTION AND THREATS

Levels of protection of perhumid coastal and montane forests vary from province to province in Eastern Canada (see table 10-1 for an overall estimate for the region using the World Protected Areas Database).

Nova Scotia

Nearly 9 percent of Nova Scotia (all ecosystem types) is protected in provincial and national parks, nature reserves, and wilderness areas, the highest proportion of any province in the region (Freedman et al. 2010). However, it also has a long history of forest clearance or alteration by logging, burning, and other disturbances (Davis and Browne 1998). The Tangier Grand Lake Wilderness Area (16,000 hectares) is the largest protected area that includes perhumid, near-coastal, fir-dominated forests in mainland Nova Scotia. Relatively little-disturbed by logging, its forests have well-developed, species-rich communities of epiphytic lichens, including several locations for *Erioderma pedicellatum* (Cameron and Richardson 2006). Cape Breton Highlands National Park (95,000 hectares) includes a significant representation of montane balsam fir forests with populations of *Platismatia norvegica*, among other boreal oceanic lichens. Old-growth, perhumid, hemiboreal, red spruce–dominated, and mixed forests are protected in several smaller areas, such as the Panuke Lake Nature Reserve and Cape Chignecto Provincial Park.

New Brunswick

The Fundy Coastal region has the largest proportional area under protection of any of New Brunswick's ecoregions (Zelazny 2007). Fundy National Park (21,000 hectares) and several areas smaller than 1,000 hectares (e.g., Little Salmon River Protected Natural Area, Herring Cove Provincial Park) have old-growth, red spruce–dominated stands. Wet, coastal fir forests are uncommon and under-protected; examples can be found in New River Beach Provincial Park and Roosevelt-Campobello International Park. The latter includes the locality where *Erioderma pedicellatum* was originally discovered in 1902. However, this species has not been found subsequently in New Brunswick (see box 4-1). Wet fir-dominated and other forests on Grand Manan, the largest island (14,000 hectares) in the outer Bay of Fundy, have been greatly altered by logging and other disturbances. No forests here are publicly owned or protected, though they are known to harbour rare oceanic lichens. In montane north-central New Brunswick, the windward, western, high-precipitation areas (e.g., the Nalaisk Mountain–Vandine Brook area) support perhumid fir forests, but these are not yet adequately captured in protected areas.

Newfoundland and Labrador

The island of Newfoundland has the only representation in eastern North America of distinctly boreal, perhumid, and (or) oceanic forests. However, only a small fraction of these communities is protected. Wet, fir-dominated forests in the central Avalon Peninsula and southern Newfoundland harbour over 95 percent of the global population of *Erioderma pedicellatum*. Many other specialized, as-yet-unrecorded species of fungi, invertebrates, and microorganisms may also have their ecological optima here. In the Avalon Forest Ecoregion (Damman 1983), the mesic and wet-fir dominated forests have been heavily impacted by clear-cut logging. Long-standing proposals to protect the most significant remaining lichen-rich fir forests in the Ripple Pond, Halls Gullies, and Lockyers Waters areas of this ecoregion have yet to be implemented.

Browsing of regenerating fir by moose (*Alces alces*) is a critical problem for many of Newfoundland's fir forests. Moose are not native to the island, but from only five animals introduced in 1878 and 1904, and in the absence of predation by wolves (*Canis lupus*), their population has exploded to more than 150,000 (McLaren et al. 2004).

The two largest protected areas in the province are the Avalon and Bay du Nord Wilderness Reserves (396,500 hectares) and Gros Morne and Terra Nova National Parks (220,500 hectares). These include relatively little of the perhumid balsam fir community typified by the occurrence of *Erioderma pedicellatum*. Protection of exceptionally old, slow-growing, montane fir-dominated forests in a portion of the Main River watershed of western Newfoundland has been achieved recently through creation of a 20,000-hectare Provincial Waterway Park. These stands have no known history of fire and, more remarkably, no record of spruce budworm or other insect epidemics (McCarthy and Weetman 2006), making them unique even among old, perhumid, balsam fir forests. The oldest-known balsam fir trees in the world, more than 250 years in age, have been found in this area (McCarthy and Weetman 2006).

Quebec

Quebec's perhumid forests are found mainly in island-like, boreal-montane ("oroboreal") areas. There are good examples in the Parc national de la Gaspésie and in the Laurentian highlands north of Quebec City. In the latter area, the Parc national des Grands-Jardins (31,000 hectares) has disjunct black spruce-lichen woodlands that may be the wettest variant of this vegetation type occurring anywhere in North America. The adjoining Réserve faunique des Laurentides and Laval University's Forêt Montmorency have significant perhumid fir forests, but these are not protected from logging. In the Appalachian region of southern Quebec, hemiboreal to boreal, montane red spruce– and fir-dominated forests are protected, for example, in the Parc national du Mont-Mégantic, and in the Réserve naturelle des Montagnes-Vertes (www.rnmv.ca). The proposed Réserve écologique du Mont-Gosford also takes in perhumid Appalachian montane forests in this region.

The least well known and possibly the most threatened of Quebec's perhumid forests are boreal spruce-fir communities occurring in the wettest portions of the hinterlands north of the Gulf of St. Lawrence. Seven of the 11 largest rivers in this region have major existing or planned hydroelectric dams—projects made possible by the perhumid climate of what is otherwise an undeveloped wilderness region. A number of studies have examined forest dynamics in areas northwest of Anticosti Island. However, if the map of moisture regions (Phillips et al. 1990; see figure 4-2) is reliable even in its general outlines, the most perhumid forests may be situated in the largely uninvestigated area between the Magpie and Little Mecatina Rivers.

CONSERVATION PRIORITIES

Looming beyond specific local threats to Eastern Canadian perhumid forests is the broader one of rapid ecosystem change caused by climate change. A range of climate models for Quebec (Bourque and Simonet 2008) and Atlantic Canada (Vasseur and Catto 2008) predict mean-annual-temperature increases of about 2°C by 2050 and 4°C by 2080. Smaller increases are forecast for coastal areas where cold-water currents and strong tidal action provide more buffering of temperatures. Precipitation amounts are expected to increase across the region but may be insufficient to offset the increased evapotranspiration resulting from warmer temperatures (Vasseur and Catto 2008; Mohan et al. 2009). If so, increases in fire frequency, insect epidemics, and decomposition rates could ensue, releasing carbon reserves from old forests and from the region's extensive peatlands (see Carlson et al. 2009).

Even if perhumid climatic conditions persist in some coastal areas of Eastern Canada, northward shifts in the ranges of species and communities are to be expected. Interactions with non–climatic variables during such shifts are difficult to predict (Mohan et al. 2009). A particular concern may be the lack of a perhumid migrational corridor linking the southern and northern perhumid areas of the region. The Gulf of St. Lawrence has posed a barrier historically to the northeastward dispersal of temperate species from Nova Scotia to the island of Newfoundland (Damman 1983; Clayden 2010). Under climate-warming scenarios, this effect will probably persist or be reinforced for species that reach their northeastern range limits in the perhumid forests of Nova Scotia, and that have limited dispersability.

Perhumid montane spruce-fir forests in the region are likely to be shifted higher in elevation. Already under stress from acidic cloud-water deposition and from more-frequent winter thaws (Mohan et al. 2009), those occupying narrow elevational belts near mountain summits could be eliminated. There is also evidence that the average elevation of the cloud base over the Appalachian Mountains is increasing in response to warming temperatures (Richardson et al. 2003). If continued, this could reduce cloud-water inputs to the water budgets of montane forests, causing the loss of their perhumid character and associated biota (Clayden 2006).

Regarding these issues, and many others relating to the ecology and biodiversity of perhumid forests in Eastern Canada, much remains to be learned. Viewing at least some of these communities as rainforests, as we have proposed here, places them in a context that we hope will draw attention to their unusual features, and to the need for research and conservation efforts focused on them. Not all temperate and boreal rainforests, no more than all old-growth forests, share the massive structure and great age emblematic of their best-known expressions in other regions of the world. Those of Eastern Canada are constituted and structured in subtler, though scarcely less interesting ways.

LITERATURE CITED

Ahti, T. 1964. Macrolichens and their zonal distribution in boreal and arctic Ontario, Canada. *Annales Botanici Fennici* 1:1–35.

———. 1980. Definition and subdivision of the subarctic: A circumpolar view. *Canadian Botanical Association Bulletin*, Supplement 13 (2):3–10.

———. 1983. Lichens. Pp. 319–60 in *Biogeography and ecology of the island of Newfoundland*. Ed. G. R. South. The Hague: Dr. W. Junk Publishers.

Arvidsson, L. 1991. Lichenological studies in Ecuador. Pp. 123–34 in *Tropical lichens: Their systematics, conservation, and ecology*. Ed. D. J. Galloway. Oxford: Oxford Univ. Press.

Bouchard, M., D. Pothier, and S. Gauthier. 2008. Fire return intervals and tree species succession in the North Shore region of eastern Quebec. *Canadian Journal of Forest Research* 38:1621–33.

Bourque, A., and G. Simonet. 2008. Quebec. Pp. 171–226 in *From impacts to adaptation: Canada in a changing climate 2007.* Ed. D. S. Lemmen, F. J. Warren, J. Lacroix, and E. Bush. Ottawa: Government of Canada.

Cameron, R. P., and D. H. S. Richardson. 2006. Occurrence and abundance of epiphytic cyanolichens in protected areas of Nova Scotia, Canada. *Opuscula Philolichenum* 3:5–14.

———, and T. Neily. 2008. Heuristic model for identifying the habitats of *Erioderma pedicellatum* and other rare cyanolichens in Nova Scotia, Canada. *Bryologist* 111:650–58.

———, T. Neily, S. R. Clayden, and W. S. G. Maass. 2009. *COSEWIC assessment and status report on vole ears,* Erioderma mollissimum, *in Canada.* Ottawa: Committee on the Status of Endangered Wildlife in Canada.

Carlson, M., J. Wells, and D. Roberts. 2009. *The carbon the world forgot: Conserving the capacity of Canada's boreal forest region to mitigate and adapt to climate change.* Seattle and Ottawa: Boreal Songbird Initiative and Canadian Boreal Initiative.

Clayden, S. R. 2000. History, physical setting, and regional variation of the flora. Pp. 35–73 in *Flora of New Brunswick.* 2d ed., by H. R. Hinds. Fredericton, NB: Department of Biology, University of New Brunswick.

———. 2006. *COSEWIC assessment and status report on ghost antler,* Pseudevernia cladonia, *in Canada.* Ottawa: Committee on the Status of Endangered Wildlife in Canada.

———. 2010. Lichens and allied fungi of the Atlantic Maritime Ecozone. In *Assessment of species diversity in the Atlantic Maritime Ecozone.* Ed. D. F. McAlpine and I. M. Smith. Ottawa: NRC Research Press. Forthcoming.

Cogbill, C. V., and P. S. White. 1991. The latitude-elevation relationship for spruce-fir forest and treeline along the Appalachian mountain chain. *Vegetatio* 94:153–75.

Cox, R., G. Lemieux, and M. Lodin. 1996. The assessment and condition of Fundy white birches in relation to ambient exposure to acid marine fogs. *Canadian Journal of Forest Research* 26:682–88.

Dahl, E. 1998. *The phytogeography of northern Europe.* Cambridge, UK: Cambridge Univ. Press.

Damman, A. W. H. 1983. An ecological subdivision of the island of Newfoundland. Pp. 163–206 in *Biogeography and ecology of the island of Newfoundland.* Ed. G. R. South. The Hague: Dr. W. Junk Publishers.

Davis, D. S., and S. Browne. 1998. *The natural history of Nova Scotia.* Halifax: Nova Scotia Museum and Nimbus Publishing.

Desponts, M., A. Desrochers, L. Bélanger, and J. Huot. 2002. Structure de sapinières aménagées et anciennes du massif des Laurentides (Québec) et diversité des plantes invasculaires. *Canadian Journal of Forest Research* 32:2077–93.

———, G. Brunet, L. Bélanger, and M. Bouchard. 2004. The eastern boreal old-growth balsam fir forest: A distinct ecosystem. *Canadian Journal of Botany* 82:830–49.

Environment Canada. 2002. *Canadian Climate Normals 1971–2000.* www.climate.weather office.ec.gc.ca/climate_normals/index_e.html

Feddema, J. J. 2005. A revised Thornthwaite-type global climate classification. *Physical Geography* 26:442–66.

Foster, D. R. 1983. The history and pattern of fire in the boreal forest of southeastern Labrador. *Canadian Journal of Botany* 61:2459–71.

Freedman, B., M. Morrison, R. Vanderkam, and E. Wiken. 2010. Protected area in the Atlantic Maritime Ecozone. In *Assessment of species diversity in the Atlantic Maritime Ecozone*. Ed. D. F. McAlpine and I. M. Smith. Ottawa: NRC Research Press. Forthcoming.

Gowan. S. P., and I. M. Brodo. 1988. The lichens of Fundy National Park, New Brunswick,
Canada. *Bryologist* 91:255–25.

Goward, T., and T. Ahti. 1992. Macrolichens and their zonal distribution in Wells Gray
Provincial Park and its vicinity, British Columbia, Canada. *Acta Botanica Fennica* 147:1–
60.

———, and T. Spribille. 2005. Lichenological evidence for the recognition of inland rainforests in western North America. *Journal of Biogeography* 32:1209–19.

Graham, A. 1999. *Late Cretaceous and Cenozoic history of North American vegetation*. New York:
Oxford Univ. Press.

Gray, J. T., and R. J. E. Brown. 1979. Permafrost existence and distribution in the Chic-
Chocs Mountains, Gaspésie, Québec. *Géographie physique et Quaternaire* 33:299–316.

Hämet-Ahti, L. 1981. The boreal zone and its biotic subdivision. *Fennia* 159:69–75.

Hare, F. K. 1961. *The restless atmosphere*. London: Hutchinson Univ. Library.

———, and M. K. Thomas. 1974. *Climate Canada*. Toronto: Wiley.

Holien, H., and T. Tønsberg. 1996. Boreal regnskog i Norge: Habitatet for trøndelagsele-
mentets lavarter. *Blyttia* 54:157–77.

Lauzon, E., D. Kneeshaw, and Y. Bergeron. 2007. Reconstruction of fire history (1680–
2003) in Gaspesian mixed-wood boreal forests of Eastern Canada. *Forest Ecology and
Management* 244:41–49.

Loucks, O. L. 1962. A forest classification for the Maritime Provinces. *Proceedings of the Scot-
ian Institute of Science* 25:85–167.

Maass, W. S. G., and D. Yetman. 2002. COSEWIC assessment and status report on the boreal
felt lichen, *Erioderma pedicellatum*, in Canada. Ottawa: Committee on the Status of Endangered Wildlife in Canada.

Marcotte, G., and M. M. Grandtner. 1974. *Étude écologique de la végétation forestière du Mont
Mégantic*. Gouvernement du Québec, Ministère des terres et forêts, Direction générale
des forêts, Service de la recherche, Mémoire 19:1–156.

McCarthy, J. W. 2001. Gap dynamics of forest trees: A review with particular attention to
boreal forests. *Environmental Reviews* 9:1–59.

———, and G. Weetman. 2006. Age and size structure of gap-dynamic, old-growth boreal
forest stands in Newfoundland. *Silva Fennica* 40:209–30.

McLaren, B. E., B. A. Roberts, N. Djan-Chékar, and K. P. Lewis. 2004. Effects of overabun-
dant moose on the Newfoundland landscape. *Alces* 40:45–59.

Mohan, J. E., R. M. Cox, and L. R. Iverson. 2009. Composition and carbon dynamics of
forests in northeastern North America in a future, warmer world. *Canadian Journal of
Forest Research* 39:213–30.

Mohnen, V. A. 1992. Atmospheric deposition and pollutant exposure of eastern U.S. forests.
Pp. 64–124 in *Ecology and decline of red spruce in the eastern United States*. Ed. C. Eagar and
M. B. Adams. New York: Springer-Verlag.

Mosseler, A., J. A. Lynds, and J. E. Major. 2003. Old-growth forests of the Acadian Forest
Region. *Environmental Reviews* 11:S47–77.

Perron, M., D. J. Perry, C. Andalo, and J. Bousquet. 2000. Evidence from sequence-tagged-site markers of a recent progenitor-derivative species pair in conifers. *Proceedings of the National Academy of Sciences, USA* 97:11331–36.

Phillips, B., and C. P. Laroque. 2006. Historical and dendrochronological assessment of the Bay of Fundy forests for dendroclimatological modeling, New Brunswick [abstract]. Atlantic Geoscience Society Colloquium. *Atlantic Geology* 42:108.

Phillips, D. W., J. M. M. Frappier, and D. E. Richardson. 1990. Canada—climatic regions: Thornthwaite classification, moisture regions, 1:7,500,000 map. In *National Atlas of Canada*. 5th ed. Ottawa: Energy, Mines and Resources Canada.

Potter, K. M., J. Frampton, S. A. Josserand, and C. D. Nelson. 2010. Evolutionary history of two endemic Appalachian conifers revealed using microsatellite markers. *Conservation Genetics*. Forthcoming.

Reiners, W. A., and G. E. Lang. 1979. Vegetational patterns and processes in the balsam fir zone, White Mountains, New Hampshire. *Ecology* 60:403–17.

Reinhardt, K., and W. K. Smith. 2008. Leaf gas exchange of understory spruce-fir saplings in relict cloud forests, southern Appalachian Mountains, USA. *Tree Physiology* 28:113–22.

Richardson, A. D., E. G. Denny, T. G. Siccama, and X. Lee. 2003. Evidence for a rising cloud ceiling in eastern North America. *Journal of Climate* 16:2093–98.

Rowe, J. S. 1972. *Forest regions of Canada*. Department of the Environment, Canadian Forestry Service, Publication No. 1300.

Sanderson, M. 1948. The climates of Canada according to the new Thornthwaite classification. *Scientific Agriculture* 28:501–17.

Schemenauer, R. S., C. M. Banic, and N. Urquizo. 1995. High-elevation fog and precipitation chemistry in southern Quebec, Canada. *Atmospheric Environment* 29:2235–52.

Shanks, R. E. 1954. Climates of the Great Smoky Mountains. *Ecology* 35:354–61.

Thompson, I. D., D. J. Larson, and W. A. Montevecchi. 2003. Characterization of old "wet boreal" forests, with an example from balsam fir forests of western Newfoundland. *Environmental Reviews* 11:S23–46.

Thornthwaite, C. W. 1931. The climates of North America according to a new classification. *Geographical Review* 21:633–55.

———. 1948. An approach toward a rational classification of climate. *Geographical Review* 38:55–94.

Tønsberg, T. 2006. Notes on some lichens from the Great Smoky Mountains National Park, North Carolina/Tennessee. *Evansia* 23:61–63.

Tuhkanen, S. 1984. A circumboreal system of climatic-phytogeographical regions. *Acta Botanica Fennica* 127:1–50.

Vasseur, L., and N. Catto. 2008. Atlantic Canada. Pp. 119–70 in *Impacts to adaptation: Canada in a changing climate 2007*. Ed. D. S. Lemmen, F. J. Warren, J. Lacroix, and E. Bush. Ottawa: Government of Canada.

Wein, R. W., and J. M. Moore. 1977. Fire history and rotations in the New Brunswick Acadian Forest. *Canadian Journal of Forest Research* 7:285–94.

——— and ———. 1979. Fire history and recent fire rotation periods in the Nova Scotia Acadian forest. *Canadian Journal of Forest Research* 9:166–78.

White, P. S., and C. V. Cogbill. 1992. Spruce-fir forests of eastern North America. Pp. 3–39 in *Ecology and decline of red spruce in the eastern United States.* Ed. C. Eager and M. B. Adams. New York: Springer-Verlag.

Yin, X., and P. A. Arp. 1994. Fog contributions to the water budget of forested watersheds in the Canadian Maritime Provinces: a generalized algorithm for low elevations. *Atmosphere-Ocean* 32:553–66.

Zelazny, V. F., ed. 2007. *Our landscape heritage: The story of ecological land classification in New Brunswick.* Fredericton, NB: New Brunswick Department of Natural Resources.

CHAPTER 5

Valdivian Temperate Rainforests of Chile and Argentina

David Tecklin, Dominick A. DellaSala, Federico Luebert,
and Patricio Pliscoff

At first glance, the Valdivian temperate rainforest of Chile and Argentina (see figure 5-1) is a mirror image of the Pacific Coast of North America in appearance and climate, but upon closer inspection this strikingly unique rainforest is dominated by broadleaf evergreen flowering trees that evolved in near complete isolation from the temperate forests of the Northern Hemisphere. This is truly a region worthy of global recognition.

GLOBAL SUPERLATIVES

The only major tree families shared between these regions such as the cedar and cypress family (Cupressaceae), are in fact those that evolved well before the continents split apart some 200 million years ago. While dense conifer forests dominate most rainforests in the Northern Hemisphere, Valdivian temperate rainforests are structurally complex and open forests with many species unique to the region. Conifers are relatively rare and occur mostly mixed in with broadleaf species.

The splitting apart of supercontinents gave rise to unique species (endemics) with affinities as far away as Australia. Distinct assemblages have evolved from this continental drift and from the rainforest's isolation by the Pacific Ocean to the west and semi-arid environments to the north and east, leading to the characterization of southern temperate rainforests as a "biogeographic is-

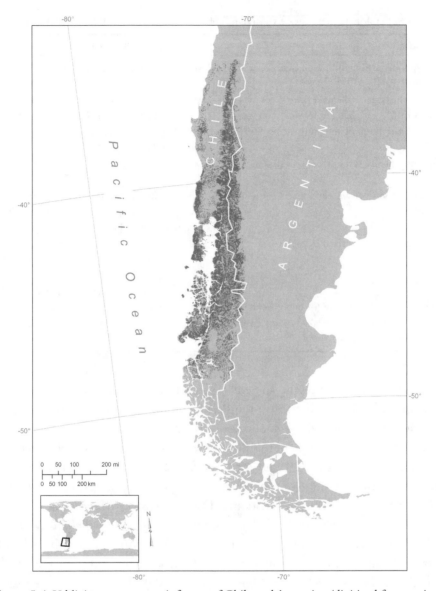

Figure 5-1. Valdivian temperate rainforests of Chile and Argentina (digitized from national vegetation surveys).

land" (Armesto et al. 1995). As with other isolated systems, the region derives much of its ecological significance from unique and varied plant assemblages (see table 5-1), as half of plant species and a third of woody plant genera are endemic (Arroyo and Hoffmann 1997). Of these genera, almost 80 percent are represented by a single species. Additionally, of the 32 genera of trees (which include 44 species), 26 (81 percent) are represented by one species. Such

Table 5-1. Global attributes of Valdivian temperate rainforests of Chile and Argentina.

Attribute	Importance
Plant affinities arising from Gondwana over 200 million years ago	Numerous adaptive radiations and rare taxa
Some of the largest, tallest, and oldest conifers in the world (e.g., alerce can live to 3,600 years)	Habitat for old-forest-dependent species (e.g., Magellanic woodpecker, rufous-legged owl)
Extraordinary levels of endemism across taxa	Globally rare, important evolutionary processes (ancient lineages)
Highly productive forests that rival Alaskan rainforests	Carbon cycling and storage, complex food-web dynamics
One of the most volcanically active areas in the world	Influences species turnover rates and provides for diversity of age classes associated with high species richness
Coastal zones characterized by high productivity	Extraordinary diversity of macroalgae (more than 60 families identified), benthic (close to or living on the bottom) invertebrates, hydrocorals, fish, and marine mammals
Enormous volumes of pure (or oligotrophic) freshwater flows through lake systems, rivers, glaciers, and ice fields	More than half of all such ice in Latin America, source of drinking water, diverse aquatic ecosystems

monotypic genera represent ancient taxa that were more widely distributed over southern South America in the Paleogene (65–23 million years ago), but survive today in the region's cool coastal climate.

But the region's uniqueness is not limited to just plants. Over 60 percent of amphibians and reptiles, 20 percent of freshwater fish and mammals, and 30 percent of birds are endemic to southern temperate rainforests (Arroyo et al. 2006). Several amphibians, including Darwin's toad (*Rhinoderma darwinii*), ground frog (*Eusophus contulmoensis*), Mehuin's frog (*Insuetrophrinus acarpicus*), and *Telmatobufo bullocki* (no common name exists for this species) have highly restricted ranges; some are limited to a single small watershed or montane area (Ortiz and Ibarra-Vidal 2005; Mendez et al. 2005). Seven of nine genera of small mammals are endemic to these rainforests (Murúa 1995), including two marsupials: the *monito del monte* ("little mountain monkey"—*Dromiciops gliroides*) and the Chilean shrew opossum (*Rhyncholestes raphanurus*). Valdivian temperate rainforests also are rich in canopy dwelling (arboreal) species, in part, because of the different combinations of forest types produced by distinct environmental gradients (Luebert and Pliscoff 2006). In addition, both austral and altitudinal migratory birds make their homes in these rainforests, including species dependent on older forest conditions, such as the giant Magellanic woodpecker (*Campephilus magellanicus*) and the rufous-legged owl (*Strix rufipes*) (Fjeldsa and Krabbe 1990).

The immense flow of water through this region is one of its key characteristics and most scenic features. Also diagnostic are large lakes of extremely pure (oligotrophic) water (Soto and Campos 1996), hundreds of mostly short river systems flowing into the Pacific, and glaciers at the tip of the Southern Continental Ice Fields. Hydrologic flows connect forested watersheds closely with the dozens of fjords that until recently were highly pristine and composed of diverse estuarine and coastal habitats (Hucke et al. 2006; Häussermann and Førsterra 2007).

The Valdivian temperate rainforest was little known to international conservationists until the 1980s. Since then, and following the growth and spread of ecological information on the region, most major conservation organizations have recognized its global significance (see table 5-1). The World Wildlife Fund (WWF) included the region as one of 200 top "ecoregions" (large regions that are defined by similarities in landform, climate, and broad species assemblages—Olson and Dinerstein 1998) in the world, requiring immediate conservation action. The northern half of the ecoregion has been identified by Conservation International as one of the world's 34 "biodiversity hotspots" (areas where the richness of life is especially concentrated and threatened; see Myers et al. 2000). The World Resources Institute (WRI) categorized much of the region's southern extent as "frontier forests" (Bryant et al. 1997) for their status as large, intact, natural forests. Additional international recognitions include designation as a Center of Plant Diversity, according to a study by WWF and the World Conservation Union (or IUCN—Arroyo and Hoffman 1997), and the presence of global endemic bird areas recognized by Birdlife International (Stattersfield et al. 1998).

ANCIENT GONDWANA AND PANGAEA ORIGINS

An unusually high proportion of plant species in this ecoregion have lineages dating back millions of years (also see Australasia, chapter 8). The most ancient elements of the Valdivian woody plants have their origin in the Jurassic period (200–144 million years ago) or earlier, as has been documented in the fossil record (Torres and Philippe 2002). At the time, Chile formed part of the supercontinent Pangaea, which contained all of the world's landmasses in a single massive island surrounded by a global sea. When the continents began to split apart, around 200 million years ago, some of the landmasses banded together to form a second supercontinent in the southern hemisphere that became Gondwanaland: an ancient land out of which arose many of Chile's present plants. Gondwana then split apart during the Cretaceous (~140 million years ago), eventually becoming Antarctica, South America, Africa, India, and Australia.

The elements of this ancient flora are still conspicuous today (Villagrán and Hinojosa 1997), but were dominant as part of the Tertiary flora of southern South America as long as 65 million years ago (Axelrod et al. 1991). Species of Neotropical and Gondwana origins have over time further increased the region's richness. In addition, many temperate or subalpine species have colonized and evolved from North America since the two continents were connected by land emergence in Central America a few million years ago. Most of these species have been able to disperse along the Cordillera from the Rockies and Sierra Madres south to the Andes. Several species, such as coastal strawberry (*Fragaria chiloensis*) and several coastal grass species, only exist in high-latitude rainforest regions in North America and along the Chilean coast, suggesting long distance dispersal (e.g., migratory shorebirds).

The current flora of Valdivia's temperate rainforests is therefore a biogeographic legacy of Gondwana origins that combined with Tertiary geologic and climatic events and glacial pulsations during the Quaternary period (1.8 million years ago). Approximately 450 species of vascular plants, representing 205 genera, are "housed" in the ecoregion, with these kinds of ancient affinities. An astonishing third of the 82 genera of woody plants are old enough to be of southern Gondwana origin, with nearest relatives (extant or fossil) in Australia, New Zealand, New Caledonia, or New Guinea. Another 25 percent are considered to have more-recent Neotropical affinities.

During glacial periods, plants needing warmer climates took refuge in the Chilean Coastal Range, where glacial and permafrost effects were less intense. A northwest displacement of the main vegetation units has been shown for the Late Glacial Maximum, which occurred 18,000 years ago in south-central Chile with a recolonization during the postglacial period (Villagrán 2001). As a consequence, the richness and uniqueness (endemism) of rainforest species tend to be higher in the glacial refugia of the Coastal Range than in the Andes (Smith-Ramírez 2005).

Human habitation of the southern temperate forests is thought to have begun some 12,000 years ago, and the archeological record shows a long history of adaptation and diversification of human societies (Dillehay 1989, 2004). At the time of European settlement, which accelerated in the second half of the nineteenth century, the region was occupied by multiple indigenous peoples—

in particular, the Mapuche people, whose hunting, gathering, horticultural, and later pastoralist lifestyle meant they were thoroughly interconnected with forest and coastal ecosystems (Navarro and Pino 1999; Adan et al. 2001). There are now approximately 3,000 rural Mapuche communities in the region, and these are estimated to hold tenure over nearly 300,000 hectares of forest, including several of the intact forest blocks considered crucial for conservation (Tecklin and Catalan 2005). One emblematic example is the Pehuenche people whose diet and ritual calendar centers on the seeds of the Araucaria tree.

WHAT AND WHERE IS THE VALDIVIAN TEMPERATE RAINFOREST?

There is currently no consensus regarding the precise climatic or vegetative limits of these remarkable rainforests, as diverse plant communities are found in a wide array of climates (Arroyo et al. 1995). While there are different classification systems for the Valdivian ecoregion (see below), in this chapter we have delineated the region as most of the Chilean territory and adjacent Argentina between 36 and 47°S latitude (see fig. 5.1). The Valdivian ecoregion includes forests dominated primarily by southern beach (*Nothofagus* spp.), evergreen microphyllous (small-leaved), and lauriphyllous (hard-leaved) forests as well as conifers. The region as a whole is approximately 34.6 million hectares, and it is estimated that, historically, 21 million hectares of these were covered in forest (Lara et al. 2000), though today this forest cover has been reduced by 40 percent to 12.7 million hectares.

From here, the southern temperate rainforest continues southwards through the Magellanic region and the island of Tierra del Fuego. However, these Magellanic or subpolar forests are generally classified as a distinct ecoregion (Dinerstein et al. 1995) and are not included in the discussion of Valdivian temperate rainforests. The Magellanic forests are subject to colder and more-humid conditions without summer drought, and overlap floristically and structurally with the forests in the southernmost portion of the Valdivian ecoregion (e.g., *N. nitida* forest), but are dominated primarily by Coihue de Magallanes (*N. betuloides*) and by lenga (*N. pumilio*) at higher elevations and on Tierra del Fuego. They tend to be structurally simpler and less species rich, though certain groups such as epiphytes are exceptionally abundant and diverse. In general, this region has seen much less human impact, has a high level of official protection, and lower development pressures than the Valdivian rainforest.

The Valdivian rainforest reaches its northern limit at the intersection with the Chilean Matorral ecoregion, which typically comprises sclerophyllous (hard-leaved) woodlands and thorny shrublands adapted to a Mediterranean-type climate. Some of these same elements are present in the large transition zone of *Nothofagus*-deciduous forests of the northern portion of the Valdivian ecoregion, but are absent south of 40°S latitude, where the last sclerophyllous remnant is found. The Valdivian rainforest is isolated from other forests of South America by the Andean Cordillera and the dry zones of the so called "South American Arid Diagonal" to the east, which includes the Patagonian steppe (a broad plain without trees), the Monte shrublands, and the Atacama and Peruvian Deserts (the driest places on earth) to the north. The eastern side of the ecoregion, where the montane cypress (*Austrocedrus chilensis*) forests occur, forms a gradient from forest to steppe (Kitzberger et al. 1997). While the limits of this region are broadly defined, the temperate rainforest itself is limited to lower and middle elevations that meet climatic criteria of temperate rainforest (e.g., Alaback 1991; Amigo and Ramírez 1998; see chapter 1).

RAINFOREST GRADIENTS AND PRODUCTIVITY

The Valdivian ecoregion is made up of four large landforms: a low Coastal Range that is the oldest geologically and enjoys the most equable climate; Andean foothills, or pre-cordillera, where the large lake systems are found along with distinct riparian vegetation; the Andes proper, reaching altitudes of nearly 3,000 meters, with the greatest variability in relief; and Central Depression, a relatively flat zone of tectonic origin lying between the two ranges. Like coastal rainforests of North America (see chapter 2), Valdivia's rainforests are influenced by the mixing of species and ecological processes through a close association of terrestrial, marine, and freshwater systems. The primary productivity in terrestrial systems is associated with latitudinal gradients where native forest was originally most abundant in low-elevation zones. This gradient is repeated in the richness of plants, which decreases both with elevation and latitude, partially as a result of the intensity of Pleistocene glaciations in those areas. In contrast to many regions in the Northern Hemisphere, however, there are several species that only occur at mid to high latitudes, since they originated from Gondwana.

The richness of freshwater fish species shows a remarkable latitudinal break at approximately 42°S latitude, below which the number of species drops sharply. However, the fjords and channels and inner sea area generate distinct conditions for marine biota that help to explain the rich array of marine life present (Vila et al. 2006).

SUBREGIONAL RAINFOREST CLASSIFICATIONS AND CLIMATE

Ecologists, geographers, and botanists have classified the Valdivian temperate rainforests according to different criteria applied across spatial scales. At the regional scale, most of the classifications have taken into account floristic diversity, especially the biogeographic zoning of plants. For the purposes of this chapter we will limit discussion to climatic zonations and specific forest types.

Climatic Zonation

Along the coast, mean annual rainfall increases and temperature decreases generally along a north–south gradient from seasonal and warm temperate (10–15°C in the warmest months, less than 2,000 millimeters annual precipitation) to perhumid (8°C, 4,500 millimeters) to subpolar just south of the Taitao Peninsula (47°S; Veblen and Alaback 1996). Notably, at midlatitudes (south of 42°S) the cool summer rainy season combines with high wind exposure to give the vegetation a structure and composition more similar to subpolar forests but with a distinctly cool, perhumid climate. Warm oceanic currents and the small land mass ameliorate winter temperatures here, and these forests also have been referred to as Patagonia rainforest (Veblen and Alaback 1996).

In addition to the strong rain-shadow effect of both the Coastal and Andes mountains, this has led to considerable climatic differentiation across the region. Mean annual rainfall on the western slope of the Coastal Mountains can be as much as twice the precipitation on the eastern slopes at the same latitude. The characteristic U-shape in the northern distribution of forests (which extend farthest north in the Coastal Range and Andes) is a consequence of both a north-to-south increase in precipitation and the rain-shadow effect generated by the mountains. In particular, the far northern Mediterranean-type climate is strongly seasonal with rainfall concentrated in the winter and a period of at least two consecutive months of summer "drought." Small-leaved deciduous forests dominate these high latitudes as well as upper elevations. In contrast, the perhumid to subpolar climate southward and at lower elevations supports mainly large-leaved forests, both deciduous and evergreen.

Forest Types

The deciduous-Mediterranean *Nothofagus* (*maulino*) forests are floristically differentiated from their temperate relatives by the presence of understory elements typical of the sclerophyllous woodlands (e.g., *Cryptocarya* spp., *Peumus boldo*). These elements tend to disappear as precipitation increases and the seasonality of rainfall decreases southward. They are progressively replaced by typical lauriphyllous understory (e.g., *Aextoxicon*, *Dasyphyllum*, *Laurelia*) in the deciduous-temperate forests dominated by *Nothofagus obliqua* (roble forest in the Central Depression) and *Nothfagus alpina* (northern mixed *Nothofagus* forest in the Andean Range). The highest floristic diversity is found at the transition between these two types, particularly in the Nahuelbuta Coastal Range at 37°S latitude.

Farther south, lauriphyllous forests predominate at the lower elevations, especially in the Coastal Range, where *Nothofagus* is absent, and evergreen broadleaf trees, such as laurel (*Laureliopsis phillipiana*), lingue (*Persea lingue*), ulmo (*Eucryphia cordifolia*), olivillo (*Aextoxicon punctatum*), and tineo (*Weinmannia trichosperma*), are dominant. Many (see Ramírez 2004) consider these coastal evergreen forests to be the most representative of Valdivian temperate rainforests, due to a combination of floristic diversity and structural complexity with a multilevel canopy and abundant epiphytes. The higher elevations of the southern Coastal Range (above 300 meters) also have alerce (*Fitroya Eupressoides*) forests, but these are more broadly distributed in the Andean slopes at the same latitude, intermingled with evergreen coihue (*N. dombeyi*) forests (see plate 6).

The Andean sub-Antarctic deciduous forests are dominated by lenga (*N. pumilio*) and ñirre (*N. antarctica*) and are located in the high altitudes of the Andes up to tree line at an altitude that progressively decreases to the south as an effect of decreasing temperature poleward. In this zone, the dominant species have smaller leaves than the other deciduous *Nothofagus*. In their northern range, these forests are usually intermingled with monkey puzzle (*Araucaria araucana*, plate 7), which can live up to 2,000 years. Extensive stands of older forests persist generally above 1,000 meters elevation on both sides of the Andes and in a few relict stands in the Coastal Range. These forests have been among the most valued and studied in the region for quite some time.

The southernmost portion of the region is dominated by coihue de Chiloé (*N. nitida*) and Guaitecas cypress (*Pilgerodendron uviferum*) forests, where the floristic composition of the shrub and herb layers illustrates the close relationship with alerce forests (Amigo et al. 2004). Finally, the lower eastern slopes of the Andes are covered by montane cypress (*Austrocedrus chilensis*) forests, delineating the boundary between the Valdivian rainforest and the Patagonian steppe in the most arid part of the temperate zone.

RAINFOREST DISTURBANCE DYNAMICS

Ecological disturbances of geologic and volcanic origin play a major role in the dynamics of southern temperate forests. With the possible exception of New Guinea, nowhere else in the world has tectonic and volcanic disturbance been as important to forest dynamics (Veblen et al. 1995). The region contains at least 23 active volcanoes that have a high probability of erupting over the next 50 years. Strong earthquakes and eruptions of Andean volcanoes cause ecological change onto themselves, but they also trigger landslides, flooding, and natural fires in highland meadows and woodlands. These infrequent but large disturbances substantially alter rainforest and other communities, but many Andean plants are able to recolonize volcanic soils. The araucaria forests exemplify this disturbance dynamic as their distribution is largely limited to active volcanoes. Since lightning fires are rare, the most-common natural source of fire ignitions is volcanic activity (Veblen et al. 1995). However, particularly in Argentina, certain zones dominated by lenga and cypress are not adapted to fire and do not recover well after burns. Prior to European settlement, fires were infrequent and intense. The increased frequency and intensity of human-caused fires in the drier eastern subregions, however, has created a serious threat to the stability of rainforests (González and Veblen 2007).

In general, two patterns of forest dynamics that are associated with large-scale disturbance regimes can be distinguished in the Valdivian temperate rainforests (Veblen et al. 1981). Along the Andes, the disturbance regime is more frequent and massive than in the Coastal Range. Disturbed areas are often colonized by shade-intolerant *Nothofagus*, which become the dominant species in the forest canopy, with further development of shade-tolerants and lack of regeneration of *Nothofagus* in undisturbed areas over time. The periodicity of stand-replacing disturbances in the Andean Range allows the recolonization of areas by *Nothofagus* and its dominance in the forest canopy over time, while in the Coastal Range the absence of periodic, large-scale disturbances may be the reason why old-growth forests lack *Nothofagus*. Persistence of the forest composition and structure in the Coastal Range is probably controlled by small-scale disturbances creating forest gaps that are colonized by the less shade-tolerant species already present in the forests, such as tineo (*Weinmannia trichosperma*). Several studies have shown that some *Nothofagus* species (e.g., *N. nitida* and *N. alpina*) are also relatively shade-tolerant and can recolonize gaps in the forest canopy after small-scale disturbances (Innes 1992, Pollmann and Veblen 2004).

THREATS

In the past, conversion to agriculture and timber plantations, and the intensive logging of commercially valuable trees such as alerce and Guaitecas cypress, along with related wildfires, have accounted for the loss of an estimated 60 percent of forest cover, and the degradation of the majority of remaining forests

Table 5-2. Threats to Valdivian temperate rainforests of Chile and Argentina.

Numerous taxa threatened with extinction, including amphibians (71 percent), freshwater fishes (84 percent), reptiles (31 percent), mammals (81 percent), and birds (25 percent).

Over half of native forests logged or converted to other uses.

Rapidly expanding paper-pulp and timber industry, explosive growth of salmon aquaculture, and dozens of large hydroelectric projects.

Mineral exploration rapidly expanding.

Global climate change could mean broad changes in rainforest distribution, regional climate, and disturbance dynamics.

(Lara et al. 2000). Unfortunately, these losses have been compounded by more-recent pressures (see table 5-2) that are expected to accelerate with global climate change. Today, the unique species, ecological processes, and extensive wildlands that constitute this region are at risk unless conservation efforts are expanded rapidly. In particular, the many narrowly distributed endemic animal and plant species are vulnerable to localized habitat destruction. Threats vary on each side of the Andes, though forest loss has been much more extensive on the Chilean side, and pressures continue to be more intense there.

The key context for contemporary pressures on forest ecosystems is a sustained rapid expansion of the Chilean economy since the 1980s, with an average growth in the Gross Domestic Production of 5.5 percent from 1990–2007 (Banco Central de Chile 2009). While relatively sparsely populated, southern Chile is an important global production center for forestry, fisheries, and agricultural industries; however, this has come at the expense of multiple environmental impacts that occur in a weak regulatory setting. A basic environmental impact assessment (EIA) system was established in the mid-1990s, but the overall *laissez faire* policy framework that has reigned since the 1970s ensures the absence of substantive environmental controls, such as land-use zoning, that are taken for granted in many countries of the global North (Tecklin et al. in press).

The limitations of the EIA system are perhaps best illustrated by the billion-dollar paper-pulp mill built in 2004 by Forestal Arauco near the city of Valdivia. The mill's wastewater discharge was approved to drain directly into an internationally protected (RAMSAR) system of estuarine wetlands famous for housing the world's largest nesting population of black-necked swans (*Cygnus melanchoryphus*). Following mill start-up, these and tens of thousands of other waterfowl either died or emigrated from the zone, sparking a still-unresolved debate over environmental reform (UACH 2005).

This mill forms part of one of the world's lowest-cost and most-profitable paper-pulp industries; it is located in the northern rainforest zone and has established over 2.2 million hectares of exotic timber plantations, primarily of *Eucalyptus* and Monterrey orradiata pine (*Pinus radiata*) (OECD-ECLAC 2005), causing wholesale landscape transformations from the earlier mosaic of native forest and agricultural lands to an unbroken, single-species monoculture. Intensive, short-rotation plantation management drives the development of high road densities, impacts watersheds and soil productivity, and has a significant toxic footprint associated with chemical paper-pulp mills and the use of pesticides. Logging of the region's native forests continues to be extensive, if largely informal, unregulated, and carried out by small operators. The vast majority (60 percent) of logging is driven by demand for firewood, which serves as the principal source of heating in Southern Chile and the country's fifth most important source of energy (Lara et al. 2006). Consequently, southern Chile's cities suffer from severe smog caused by wood burning. The intensive agricultural systems in the Central Valley, ranging from annual crops in the northern portion to dairy and berry production in the south-central area, present a major barrier to the dispersal and migration of many rainforest species, partially isolating the Coastal from the Andean Range.

Over the last decade, freshwater, estuarine, and coastal areas have experienced severe and still poorly understood effects from a booming aquaculture industry that relies on open-net pen production of Atlantic salmon (*Salmo salar*), but is expanding into many other species. This industry contributes excessive nutrients, antibiotics, and other chemicals as well as providing a major pathway for invasive species to freshwater and marine ecosystems (Leon et al. 2007; Buschman et al. 2009; also see chapter 2), and has been shown to reduce local marine biodiversity (Soto and Norambuena 2004). Although major epidemics of viral and bacterial diseases and sea lice have slowed expansion since 2008, the industry continues to extend southwards into the Magallanes Region, and the government has proposed doubling the industry over the next decade (Boston Consulting Group 2007). The largest threat to freshwater ecosystems may be, however, the new and unprecedented boom in hydroelectric development and associated infrastructure occurring in the region (also see chapter 3). While nearly all of the rivers south of the Bío Bío basin are still free-flowing, dams are proposed or underway for most of the major watersheds, including the Baker, Pascua, Puelo, and Valdivia, as well as for dozens of smaller rivers (Hall et al. 2009). In addition, over the past five years mineral exploration has taken off and concessions have been granted for extensive areas in the Aysen region, Chile's eleventh of 15 administrative regions and one of the least-contaminated places on the planet. This could constitute a major new risk if even a small portion of the nation's enormous mining sector—until now concentrated in the arid north—were to extend southward.

Due to past biogeographic isolation, the region is vulnerable to invasion by exotic species. Invasives like the red deer (*Cervus elaphus*) and wild boar (*Sus scrofa*) have localized impacts, while rabbit (*Oryctolagus cuniculus*), mink (*Mustela vison*), and feral dogs impact larger areas and have escaped efforts at control (Iriarte et al. 2005). There are a number of important invasive plant species, too, such as several leguminous shrubs (e.g., *Cytisus scoparius*, Scots broom), lupine species and others that are particularly problematical in pine plantations and disturbed areas, as well as conifers such as Douglas-fir (*Pseudotsuga menziesii*) and radiata pine (Arroyo et al. 2000). Invasive species can also be a problem in protected areas (Pauchard and Alaback 2004). Introduction of willow (*Salix* spp.) from the Northern Hemisphere has also had important effects on the dynamics and structure of some riparian areas. However, salmonids represent by far the most important threat to aquatic systems. Unlike in North American temperate rainforests, there are no native salmonids in temperate rainforests of the Southern Hemisphere. Introductions began in the nineteenth century, accelerated with the aquaculture boom since the 1990s, and have had devastating effects since then. Salmonids both prey upon and compete for habitat with native species, which due to small body size and low predation pressures are particularly vulnerable to invasion (Vila et al. 2006).

Global climate change is expected to exacerbate these pressures on forest ecosystems, although a precise understanding of its rate and impacts is currently lacking. Projected impacts will vary considerably along the region's north-south gradient, generally including higher average air temperatures, decreased precipitation, and higher variability in climate. This will affect the timing and intensity of natural-disturbance events, such as droughts, heavy rains, and lightning. Temperature increases are expected to range from 4°C in the northern portion of the ecoregion to 2°C in the south, being higher in the summer season. Precipitation declines are expected to be most severe north of 40°S latitude, and to decrease from 40 to 60 percent in relation to current precipitation patterns (University of Chile-DGF 2006). Hydrologic regimes will be heavily altered through reductions in Andean snowpack and because of the accelerated glacial melting that already has been documented in the southern portion of the ecoregion (Rivera et al. 2006).

CONSERVATION PRIORITIES

Rainforest conservation has a long history in both Chile and Argentina, where concern for forest loss dates to the early nineteenth century, and the first official protected areas were established in 1907 and 1903, respectively (Pauchard and

Villaroel 2002). Nonetheless, the historic conservation strategies and outcomes are very different in each country. Argentine society has placed a high value on these forests, and by and large they remain in state hands. The primary activities in the region are tourism and, to a lesser extent, grazing. Approximately 40 percent of the Argentine forested portion of the ecoregion (933,348 of 2,278,204 ha; but see table 10-1 in chapter 10 for estimates involving strictly protected areas only) is under some form of official protection, ranging from the national parks system, which has significant capacity and a relatively high level of management, to provincial parks that mostly exist only on paper.

In Chile, temperate rainforests have been the site of state-sponsored colonization schemes, forest conversion to plantations, and industrial exploitation. The region is divided between private lands that make up its majority and public lands that are largely restricted to the protected area system. If all types of public protected areas—including those with very limited legal and effective protection and the best-established private areas are counted—this would cover ~49 percent of the forested portion of the ecoregion (~5.1 million hectares; but see table 10-1 in chapter 10 for strictly protected areas only). Although this represents an enormous achievement, the system also suffers from major weaknesses. The vast majority of this area (more than 90 percent) is at the extreme south latitudes or higher elevations (above 600 meters) where Valdivian temperate rainforest is generally absent. The lower-elevation, productive forests of the Coastal Range, Central Valley, and northern portion of the ecoregion are poorly represented in public protected areas. Thus, of the 55 distinct vegetation types described for the ecoregion, 30 have under 10 percent of current coverage protected and 11 are totally unprotected (Luebert and Pliscoff 2006). Moreover, this system continues to be institutionally and legally precarious as well as extremely underfunded (Tacon et al. 2006a). Little public land remains in underrepresented areas for expansion of the national protected area system, and the government has been consistently unwilling to enlarge the public domain. Thus, in the future most expansion of protected-area coverage will likely come from willing private landowners.

Public efforts over the last decade have been complemented by one of the world's most dynamic trends in private conservation. Despite the absence of a regulatory framework or incentives, diverse assortments of Chilean and international individuals and organizations have spontaneously invested in land conservation. In particular, a turning point in two of the country's largest timber controversies (the Chaihuin forest conversion scheme near Valdivia, and the Trillium logging project in Tierra del Fuego) was reached in 2003 when these controversial logging operations were settled through private land transactions facilitated by international organizations. Even prior to this, the Conservation Land Trust, an organization supported by U.S. philanthropists Doug and Kris Tompkins, has been a pioneer in this area and has been directly involved in protecting over 0.5 million hectares of land in Chile. However, most initiatives are led by small- or medium-sized landowners, and represent a creative mixing of public purposes and private benefits (e.g., recreational use—Corcuera et al. 2002). Since these have no legal status there is no official registry, but recent estimates at the national level place the number of areas at over 300, covering over 1.5 million hectares, most of which is located within the ecoregion (Sepúlveda 2004; Maldonado and Faundez 2005). Given the lack of a regulatory framework, it remains unclear exactly how well protected these lands really are or will remain in the future.

Despite having one of the world's longest and most spectacular coastlines, a glaring deficit in Chile's conservation effort is the lack of coastal and marine conservation areas, as well as legal tools for protecting freshwater systems, particularly rivers. In principle, all surface water has been privatized and is available for consumptive (irrigation) or nonconsumptive (hydroelectric) use, according to the country's water code (Bauer 2004), and the current regulatory regime penalizes environmental flows as a form of "nonuse." There are currently no legal mechanisms to permanently ensure that a river remains free-flowing, and extremely limited potential within the current framework for minimum environmental flows. New legal tools for protection of rivers and lakes will be required in order to secure high-priority watersheds from development pressures. Moreover, for the coastal and nearshore marine ecosystems, both new regulation and increased governmental coordination are required in order to establish better-integrated coastal planning and conservation. Just in the last few years the government has begun to establish the first Marine and Coastal Protected Areas, including one along the Valdivian coastline (CONAMA 2008); however, such efforts confront a tangle of overlapping jurisdictions and vested interests such that progress has been limited and slow.

In addition to protected-area approaches, significant national and international efforts have sought to promote sustainable forestry in Chile through research, forest mapping, pilot projects, and extension with landowners. Unfortunately, sustainable forestry has been hampered by an inconclusive sixteen-year debate over native forest legislation. Although a new native forest law was finally enacted in 2008, this is of limited regulatory scope and focuses primarily on new subsidies for forestry that are not yet operative on a large scale. Moreover, the timber industry has for the last thirty years centered on exotic timber plantations, a trend that has reduced demand for and commercial interest in native forests.

A landmark in reducing the industries' impact has been the spread of forestry certification. The certification system of the Forest Stewardship Council (FSC, www.fsc.org) has driven this process; the medium-sized timber corporations in Chile (primarily the subsidiaries of multinational corporations) have had their lands certified by FSC-accredited certifiers beginning in the late 1990s. The largest Chilean corporations that make up the majority of the sector initially chose to certify under an alternative, industry-based standard, but in 2009 also controversially began assessments for FSC certification. Though debate continues as to the comprehensive impact of certification, one widely recognized benefit has been the establishment of strict limits on native forest conversion.

Since the late 1990s, a new generation of conservation work has focused on community-based management, especially non-timber forest products, ecotourism, and low-impact silviculture (Catalan et al. 2005; Tacon et al. 2006b). The Huilliche indigenous territory in the southern Coastal Range and the Pehuenche indigenous territory in the Araucania forests of the Andes have been particularly active areas where new ventures aimed at sustainable forest use and indigenous community self-governance have drawn national and international attention.

As in other regions, there is no "silver bullet" or panacea for Valdivian rainforest conservation. Given the diversity of public values, communities and economic sectors in the region, as well as the patchy and heterogeneous distribution of remaining natural habitats, diverse conservation strategies are required. For the moment much of the conservation effort in Valdivia is focused on conservation of remaining intact areas, which are needed to maintain viable populations of rainforest focal species (see box 5-1). At the same time, existing ecological information and predictions for future climate change point to the imperative of landscape and regional-level conservation efforts that encompass whole environmental gradients. On the one hand, this requires a rethinking and expansion of existing biodiversity protection in public and private protected areas, and, on the other, the pursuit of distinct conservation measures with a range of communities and economic sectors in more intensively used areas.

At a minimum, the national system must be strengthened legally and institutionally, and a new system of sustainable financing must be established. To complement this, a mixture of public regulation and private voluntary

---

BOX 5-1

Focal-Species Conservation in the Valdivian Temperate Rainforest of Chile and Argentina.

In the absence of detailed information on the habitat needs of most rainforest species, one important conservation planning approach has centered on "focal species" as a proxy for large groups of species. These are typically wide-ranging (area-sensitive) species with specialized diets or breeding needs, requiring large, interconnected areas to maintain viable populations (Lambeck 1997). Fifty taxa have been discussed as possible focal species for the entire Valdivian ecoregion, but only five were selected by conservation groups for planning purposes (WWF et al. 2002). They include Magellanic woodpecker, rufous-legged owl, Patagonian huemul (*Hippocamelus bisulcus*), southern river otter (*Lutra provocax*), and guigna cat (*Oncifelis guigna*). Each of these focal species uses large and contiguous areas during some portion of their annual life cycle, and several depend on older forests. They are also highly regarded as charismatic symbols of rainforest biodiversity. Based on these needs, the analysis estimated that intact blocks of forest at least 10,000 hectares should be conserved across the ecoregion.

---

mechanisms, such as certification, must be established to ensure that privately protected areas actually conserve lands over the long-term and that forestry impacts are minimized. Despite their rapid growth, community-based efforts require much stronger governmental support in land titling and political recognition, as well as technical and financial incentives for long-term management.

The Valdivian region has witnessed the coming and going of dinosaurs, the parting of ancient continents, reoccurring glaciations, and untold volcanic eruptions. Yet in the face of each of these natural calamities it has managed to sustain its remarkable web of life. Unless appropriate conservation measures are significantly expanded, however, the fate of these forests will remain highly uncertain as the climate conducive to rainforest communities shifts and land-use pressures intensify. Until then, the Valdivian temperate rainforests will take their place alongside all the world's temperate rainforests as highly threatened and in need of stepped-up protection and scientifically rigorous conservation and monitoring strategies.

LITERATURE CITED

Adan, L., R. Mera, M. Becerra, and M. Godoy. 2001. Ocupación arcaica en territorios boscosas y lacustres de la Región Precordillerana Andina (IX y X Regiones): El sitio Marifilo 1 de la localidad de Pucura. Pp. 1121–36 in *Acts of the XV National Archeological Congress* 2. Arica, Chile.

Alaback, P. B. 1991. Comparative ecology of temperate rainforests of the Americas along analogous climatic gradients. *Revista Chilena de Historia Natural* 64:399–412.

Amigo, J., and C. Ramírez. 1998. A bioclimatic classification of Chile: woodland communities in the temperate zone. *Plant Ecology* 136:9–26.

Amigo, J., C. Ramírez, and L.G. Quintanilla. 2004. The *Nothofagus nitida* (Phil.) Krasser woodlands of southern Chile in the northern half of their range: Phytosociological position. *Acta Botanica Gallica* 151:3–31.

Armesto, J., C.Villagrán, and M. T. K. Arroyo. 1995. *Ecología de los bosques nativos de Chile.* Santiago: Editorial Universitaria.

Arroyo, M. K., L. Cavieres, A. Peñaloza, M. Riveros, and A. M. Faggi. 1995. Relaciones fitogeográficas y patrones regionales de riqueza de especies en la flora del bosque lluvioso templado de Sudamérica. Pp. 71–99 in *Ecología de los bosques nativos de Chile.* Ed. Armesto, J., C.Villagrán, and M. T. K. Arroyo. Santiago: Editorial Universitaria.

———, and A. Hoffmann. 1997. Temperate rain forests of Chile. Pp. 542–48 in *Centers of plant diversity: A guide and strategy for their conservation.* Ed. S. D. Davis, V. H. Heywood, O. Herrera-MacBride, J. Villalobos, and A.C. Hamilton. Cambridge: International Union for the Conservation of Nature.

———, C. Marticorena, O. Matthei, and L. Caviares. 2000. Plant invasions in Chile: Present patterns, future predictions. Pp. 385–421 in *Invasive species in a changing world.* Ed. H.A. Mooney and R.J. Hobbs. Washington, DC: Island Press.

———, P. Marquet, C. Marticorena, J.A. Simonetti, L. Cavieres, F.A. Squeo, R. Rozzi, and F. Massardo, eds. 2006. Pp. 94–99 in *El Hotspot chileno, prioridad mundial para la conservación: Biodiversidad de Chile.* Santiago, Chile: CONAMA.

Axelrod, D. I., M. T. K. Arroyo, and P. H. Raven. 1991. Historical development of temperate vegetation in the Americas. *Revista Chilena de Historia Natural* 64:413–46.

Banco Central de Chile. 2009. The Chilean Economy at a Glance: January 2009. Santiago, Chile. www.bcentral.cl/.

Bauer, C. J. 2004. *Siren song: Chilean water law as a model for international reform.* Washington, DC: Resources for the Future.

Boston Consulting Group. 2007. Estudios de competitividad en clusters de la economía chilena: Documento de referencia Acuicultura. Boston Consulting Group—CORFO.

Bryant, D., D. Nielsen, and L. Tangley. 1997. *Last frontier forests: Ecosystems and economies on the edge.* Washington, DC: World Resources Institute.

Buschmann, A. H., F. Cabello, K. Young, J. Carvajal, D. A. Varela, and L. Henríquez. 2009. Salmon aquaculture and coastal ecosystem health in Chile: Analysis of regulations, environmental impacts and bioremediation systems. *Ocean and Coastal Management* 52:243–49.

CONAMA. 2008. *Biodiversidad de Chile: Patrimonio y desafíos.* Santiago: CONAMA and Ocho Ojos.

Corcuera, E., C. Sepúlveda, and G. Geisse. 2002. Conserving land privately: Spontaneous markets for land conservation in Chile. Pp. 127–50 in *Selling forest environmental services: Market-based mechanisms for conservation and development.* Ed. S. Paggiola, J. Bishop, and N. Landell-Mills. London: IIED-WB.

Catalan, R., P. Wilken, A. Kandzior, D. Tecklin, and H. Burschel. 2005. *Bosques y Comunidades del Sur de Chile.* Santiago: Editorial Universitaria.

Dillehay, T. 1989. *Monte Verde: A late pleistocene settlement in Chile.* Vol. 1, *Paleoenvironment and site context.* Washington, DC: Smithsonian Institution Press.

———, 2004. *Monte Verde: Un asentamiento humano del pleistoceno tardío en el sur de Chile.* Santiago: LOM Ediciones.

Dinerstein, E., D. Olson, D. Graham, A. Webster, S. Primm, M. Bookbinder, and G. Ledec. 1995. *A conservation assessment of the terrestrial ecoregions of Latin America and the Caribbean.* The International Bank for Reconstruction and Development. Washington, DC: The World Bank.

Fjeldså, J., and N. Krabbe. 1990. *Birds of the High Andes.* Copenhagen and Stenstrup, Denmark: Zoological Museum and Apollo Books.

González, M. E., and T. T. Veblen. 2007. Incendios en bosques de *Araucaria araucana* y consideraciones ecológicas al madereo de aprovechamiento en áreas recientemente quemadas. *Revista Chilena de Historia Natural* 80:243–53.

Hall, S. F., R. Roman, F. Cuevas, P. Sanchez. 2009. *Se necesitan represas en la Patagonia?* Santiago: Ocho Libros Editores.

Häussermann, V., and G. Försterra. 2007. Extraordinary abundance of hydrocorals (Cnidaria, Hydrozoa, Stylasteridae) in shallow water of the Patagonian fjord region. *Polar Biology* 30:487–92.

Hucke, R., F. Viddi, and M. Bello. 2006. Marine conservation in southern Chile: The importance of the Chiloe-Corcovado region for blue whales, biological diversity and sustainable development. Valdivia: Blue Whale Center.

Innes, J. 1992. Structure of evergreen temperate rain forest on the Taitao Peninsula, southern Chile. *Journal of Biogeography* 19:555–62.

Iriarte, A., G. A. Lobos, and F. M. Jaksic. 2005. Invasive vertebrate species in Chile and their control and monitoring by governmental agencies. *Revista Chilena de Historia Natural* 78:143–51.

Kitzberger, T., T. Veblen, and R. Villalba. 1997. Climatic influences on fires regimes along a rain forest-to-xeric woodland gradient in northern Patagonia, Argentina. *Journal of Biogeography* 24:35–47.

Lambeck, R. J. 1997. Focal species: A multi-species umbrella for nature conservation. *Conservation Biology* 11:849–56.

Lara, A., and R. Villalba. 1993. A 3620-year temperature record from *Fitzroya cupressoides* tree-rings in southern South America. *Science* 260:1104–6.

——— et al. 2000. Cobertura de los bosques de la ecorregión valdiviana de Chile y Argentina antes de la colonización europea, 1:500,000 scale. Final Report. INTA-FVSA-UACh-WWF.

————, R. Reyes, and R. Urrutia. 2006. Bosques Nativos. Pp. 107–38 in *Informe pais: Estado del medio ambiente en Chile 2005*. Ed. Universidad de Chile. Santiago: Centro de Análisis de Políticas Públicas and Lom Ediciones.

Leon, J., D. Tecklin, S. Díaz, and A. Farías. 2007. Salmon aquaculture in Chile's southern lakes-Valdivian ecoregion: History, tendencies and environmental impacts. Valdivia: WWF Chile. www.wwf.cl/publicaciones.htm.

Luebert, F., and P. Pliscoff. 2006. Sinopsis bioclimática y vegetacional de Chile. Santiago: Editorial Universitaria.

Maldonado, V., and R. Faúndez. 2005. *Actualización base de datos cartográfica de areas silvestres protegidas privadas al nivel nacional: Informe final*. CODEFF-CONAMA, Santiago, Chile.

Mendez, M. A., E. R. Soto, F. Torres-Pérez, and A. Veloso. 2005. Anfibios y reptiles de los bosques de la Cordillera de la Costa (X Región, Chile). Pp. 441–51 in *Historia, biodiversidad y ecología de los bosques costeros de Chile*. Ed. Smith-Ramírez, C., J. J. Armesto, and C. Valdovinos. Santiago: Editorial Universitaria.

Murúa, R. 1995. Comunidades de mamíferos del bosque templado de Chile. Pp. 113–33 in *Ecología de los bosques nativos de Chile*. Ed. Smith-Ramírez, C., J. J. Armesto, and C. Valdovinos. Santiago: Editorial Universitaria.

Myers, N., R. A. Mittermeier, C. G. Mittermeier, G. A. B. da Fonseca, and J. Kent. 2000. Biodiversity hotspots for conservation priorities. *Nature* 403:853–58.

Navarro, X. and M. Pino. 1999. Ocupaciones arcaicas en la costa de Valdivia: El sitio Chan Chan 18. Pp. 65–82 in *Actas de III Jornadas de Arquelogiaí la Patagonia*. Bariloche, Argentina.

Olson, D. M., and E. Dinerstein. 1998. The global 200: A representation approach to conserving the Earth's most biologically valuable ecoregions. *Conservation Biology* 12:502–15.

OECD-ECLAC. 2005. *OECD Environmental Performance Reviews: Chile*. OECD Publishing.

Ortiz, J. C., and H. Ibarra-Vidal. 2005. Anfibios y reptiles de la cordillera de Nahuelbuta. Pp. 427–39 in *Historia, biodiversidad y ecología de los bosques costeros de Chile*. Ed. Smith-Ramírez, C., J. J. Armesto, and C. Valdovinos. Santiago: Editorial Universitaria.

Pauchard, A., and P. Villarroel. 2002. Protected areas in Chile: History, current status, and challenges. *Natural Areas Journal* 22:318–30.

Pauchard, A., and P. Alaback. 2004. Influence of elevation, land use, and landscape context on alien plant invasions along roadsides in protected areas in south-central Chile. *Conservation Biology* 18:238–48.

Pollmann, W., and T. Veblen. 2004. *Nothofagus* regeneration dynamics in south-central Chile: A test of a general model. *Ecological Monographs* 74:615–34.

Rivera, A. F., A. Rivera, and C. Rodrigo. 2006. Variaciones recientes de glaciares entre 41°S y 49°S y su relación con los cambios climáticos. *Revista Geográfica* 139:39–69 (January–June, 2006).

Sepúlveda, C. 2004. ¿Cuánto hemos avanzado en conservación privada de la biodiversidad?: El aporte de las Áreas Protegidas Privadas en perspectiva. *Ambiente y Desarrollo* 20 (1):74–79.

Smith-Ramírez, C., J. Armesto, and C. Valdovinos. 2005. *Historia, biodiversidad y ecologia de los bosques costeros de Chile*. Santiago: Editorial Universitaria.

Soto, D., and H. Campos. 1996. Los lagos oligotróficos del bosque templado húmedo del sur de Chile. Pp. 317–33 in *Historia, biodiversidad y ecología de los bosques costeros de Chile*. Ed. Smith-Ramírez, C., J. J. Armesto, and C. Valdovinos. Santiago: Editorial Universitaria.

Soto, D., and F. Norambuena. 2004. Evaluation of salmon farming effects on marine systems in the inner seas of southern Chile: a large-scale mensurative experiment. *Journal of Applied Ichthyology* 20:493–501.

Stattersfield, A. J., M. J. Crosby, A. L. Long, and D. C. Wege. 1998. *Endemic bird areas of the world: Priorities for biodiversity conservation*. Cambridge: Birdlife International.

Tacon, A., U. Fernandez, A. Wolodarsky, and E Nuñez. 2006a. *Evaluación rapida de la efectividad del manejo de las áreas protegidas de la Ecorregión Valdiviana*. WWF-CONAF. (www.wwf.cl).

———, A., J. Palma, U. Fernandez, and F. Ortega. 2006b. El Mercado de los productos forestales no madereros y la conservación de los bosques del sur de Chile y Argentina. Valdivia: WWF Chile.

Tecklin, D., C. Bauer, and M. Prieto. Forthcoming. Making environmental law for the market: The emergence, characteristics and effects of Chile's environmental regime.

———, and R. Catalan. 2005. La gestión comunitaria de los bosques nativos en el sur de Chile: Situación actual y temas en discusión. Pp. 19–39 in *Comunidades y bosques del sur de Chile*. Ed. R. Catalan, P. Wilken, A. Kandzior, D. Tecklin, and H. Burschel. Santiago: Editorial Universitaria.

Torres, T., and M. Philippe. 2002. Nuevas especies de *Agathoxylon* y *Baieroxylon* del Lías de La Ligua (Chile) con una evaluación del registro paleoxilológico en el Jurásico de Sudamérica. *Revista Geológica de Chile* 29:151–65.

UACH. 2005. *Estudio sobre origen de mortalidades y disminución poblacional de aves acuáticas en el Santuario de la Naturaleza Carlos Anwandter, en la provincia de Valdivia: Final Report*. UACH-CONAMA.

University of Chile-DGF. 2006. Estudio de la variabilidad climática en Chile para el siglo XXI. Santiago: CONAMA.

Veblen, T. T., C. Donoso, F. M. Schlegel, and B. Escobar. 1981. Forest dynamics in south-central Chile. *Journal of Biogeography* 8:211–47.

———, T. Kitzberger, B. Burns, and A. Rebertus. 1995. Perturbaciones y dinámica de regeneración en bosques del sur de Chile y Argentina. Pp. 169–98 in *Ecología de los bosques nativos de Chile*. Ed. J. Armesto, C. Villagrán, and M. T. K. Arroyo. Santiago: Editorial Universitaria.

———, and P. B. Alaback. 1996. A comparative review of forest dynamics and disturbance in the temperate rainforests in North and South America. Pp. 173–213 in *High-latitude rain forests and associated ecosystems of the west coast of the Americas: Climate, hydrology, ecology and conservation*. Ed. R. Lawford, P. Alaback, and E. R. Fuentes. *Ecological Studies* 116. Springer-Verlag.

Vila, I., A. Veloso, R. Schlatter, and C. Ramírez. 2006. *Macrófitas y vertebrados de los sistemas límnicos de Chile*. Santiago, Chile: Editorial Universitaria.

Villagrán, C., and L. F. Hinojosa. 1997. Historia de los bosques del sur de Sudamérica, II: Análisis fitogeográfico. *Revista Chilena de Historia Natural* 70:241–67.

————. 2001. Un modelo de la historia de la vegetación de la Cordillera de la Costa de Chile central-sur: La hipótesis glacial de Darwin. *Revista Chilena de Historia Natural* 74:793–803.

WWF, APN, Centro de Ecología Aplicada de Nuequén, CODEFF, Fundación Senda Darwin, FVSA, INTA, et al. 2002. Resumen Visión para la Biodiversidad de la Ecorregión de los Bosques Templados Lluviosos de Chile y Argentina. Valdivia: WWF Chile. www .wwf.cl/archivos_publicaciones/resumen_vision_biodiversidad.pdf

CHAPTER 6

Temperate and Boreal Rainforest Relicts of Europe

Dominick A. DellaSala, Paul Alaback, Anton Drescher, Håkon Holien, Toby Spribille, and Katrin Ronnenberg

European temperate rainforests are disjunctly distributed from ~45° to 69°N latitude, where they are influenced by maritime climates (see figure 6-1). Storms originating in the North Atlantic and the Mediterranean (Balkans) provide for mild winters, cool summers, and adequate precipitation to sustain rainforests throughout the year. Due to extensive deforestation, however, today's European rainforests are mere fragments of primeval rainforests. A reminder of a bygone era when rainforests flourished, they are barely hanging on as contemporary rainforest relicts (see box 6-1).

FROM FOREST PRIMEVAL TO RAINFOREST RELICT

Because very few rainforests remain throughout Europe, they have not received much attention from ecologists and therefore broadly applicable classifications are lacking. We have chosen to limit discussion of rainforest relicts to regions where there exists reliable information on extent prior to widespread human influence that also meets published rainforest definitions or the conditions established by the global rainforest model described in Chapter 1. In general, information on areas where conditions are most likely to support rainforests builds on earlier mapping (Kellogg 1992), with the improvements noted in Chapter 1, particularly the addition of Central Europe and the Northwest Balkans. Thus, this chapter is primarily arranged geographically from north to

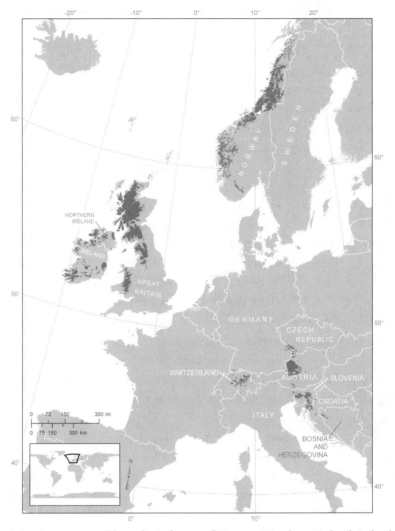

Figure 6-1. Temperate and boreal rainforests of Norway, Northern Ireland, Ireland, Great Britain, and the European Alps, based on the rainforest distribution model.

south and west to east, and summarily includes the lichen-rich rainforests of Norway (boreal and boreonemoral), and the perhumid rainforests of Ireland, Great Britain, and Central Europe and the Northwest Balkans (see table 6-1).

RAINFOREST RELICTS OF NORWAY

Norway's coastline is rimmed by mountainous fjords reminiscent of Alaska and Chile, where climatic conditions are conducive to rainforest development. Its

BOX 6-1

The Storied Rainforest Relicts of Europe.

Throughout Europe, centuries of widespread conversion of forests to agricultural pasture or grazing lands (e.g., Ireland, Britain) and deforestation for fuel wood and lumber (e.g., Norway, Ireland) resulted in destruction of nearly all forests and woodlands (Ingrouille 1995; Rackham 1995; Williams 2003; Higgins et al. 2004). European deforestation mainly occurred in several early waves of human expansion, including during the post-glacial colonization period, about 10,000 years ago, that led to the initial clearing of woodland; the Neolithic period, some 6,000 years ago, that brought about the development of organized, permanent settlements and land clearing for agriculture; and the Bronze Age, over 4,500 years ago, that accelerated deforestation through the use of metal tools (Ingrouille 1995). Additionally, the Turkish migration into the Balkans in the fifteenth and sixteenth centuries led to further losses. Since these earlier periods, invading cultures have repeatedly cleared woodland for settlement, particularly during times of population growth, political, social, and religious upheaval, and industrialization, introducing goats, pigs, cows, sheep, rabbits, and exotic plants along the way (Ingrouille 1995; Shepherd 1998; Tuite and Brown 1998; Williams 2003; Higgins et al. 2004). Consequently, the few remaining rainforest relicts are now deficient in native species, with recolonization hampered by a long history of livestock grazing, coppicing (a traditional method of woodland management in which young trees are repeatedly cut), extensive soil damage, and the insularity of Great Britain (Ingrouille 1995).

Northwest Europe, in particular, is not a species-rich place, with only 2,000 native flowering plants and few that are unique globally (Ingrouille 1995). There are several reasons for this: (1) glaciers and cold deserts have swept across the continent, periodically wiping the vegetation slate clean; (2) northern latitudes are beyond the distributional limits of many species; (3) the North Sea and Atlantic Ocean are isolating barriers, impenetrable to many colonizing plants and animals; (4) a narrow range of habitats (such as in the British Isles) limits the diversity of plant species; and (5) a long history of deforestation has wiped out nearly all woodlands (Ingrouille 1995; Williams 2003; Higgins et al. 2004).

Table 6-1. Unique attributes of temperate and boreal rainforest relicts of Europe.

Attribute	Importance
Distributed mainly as isolated fragments in Northwest and Central Europe and Northwest Balkans (~45° to ~69°N latitude)	Fragmented distribution poses challenges to conserving large landscapes
Maritime climate generates moderate to extreme precipitation delivered frequently throughout the year	Primary reason for rainforest communities
Boreal (mainly conifers), boreonemoral, elements of Caledonia pinewoods, Celtic broadleaf woodlands, and sessile oak woodlands	Restoration building blocks
High richness of mosses, ferns, and lichens	Many globally unique lichens (in Norway), several of which are "red listed" because of declining status

north-south distribution results in boreal coniferous forest in the north and boreonemoral (temperate and transitional, with broadleaf deciduous and coniferous woodland elements) forests in the south (see figure 6-2). In general, boreal forests, which also are found in high-latitude regions of Alaska, Canada, Newfoundland, and Russia, do not receive enough growing-season precipitation and have too short a growing season to qualify as rainforest. However, Norway's coastal areas have considerably longer growing seasons and more precipitation than most inland regions and therefore represent an exceptional case. Most important is the high frequency of Norway's precipitation (Alfnes and Førland 2006), which provides high humidity levels throughout most of the year. The winter climate is much more favorable for rainforest species, particularly because of the moderating influence of the North Atlantic current.

Because Norwegian rainforests are influenced by both moderate and cool climates, depending on northerly latitude, coastal boreal and boreonemoral forests fall within global definitions of subpolar (or boreal) and perhumid temperate rainforests, respectively (Alaback and Pojar 1997). A maritime climate delivers considerable moisture (more than 2,032 millimeters in mountainous areas) spread evenly during the year (Moen 1999).

Boreal Rainforests

When combined with low seasonal temperatures, high humidity, and frequent precipitation, Norway's coastal boreal forests support rainforest communities rich in lichens, mosses, and ferns, with arboreal lichens found primarily on older deciduous trees, but also locally on spruce (*Picea* spp.—Rolstad et al.

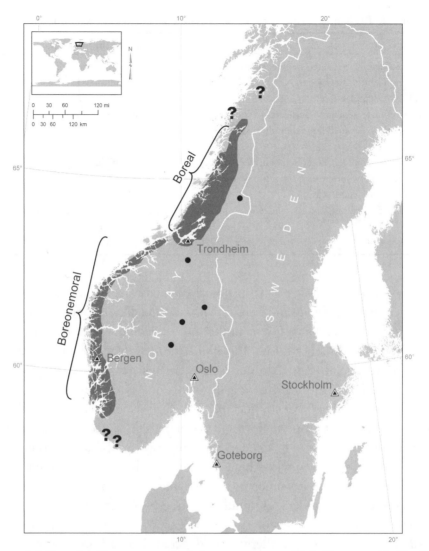

Figure 6-2. Boreal and boreonemoral rainforests of Norway (digital layer provided by Geir Gaarder, personal communication). Question marks indicate areas that have yet to be confirmed, while dots are specific locations associated with waterfalls.

2001). Less than 10 percent of these forests remain as old growth originating from natural regeneration.

Boreal rainforests of Norway are found north of the oak (*Quercus* spp.) limit to ~67°N latitude, the highest-latitude rainforests in the world; however, there are small elements as far north as 69°N latitude (and perhaps even farther north) in the valley of Målselv in Troms County. Here, rainforests generally occur on sites protected from North Atlantic storms, especially east-facing slopes and ravines with good soil drainage. Like subpolar rainforests in Alaska and southern Patagonia, trees in these forests are often much shorter (often under 30 meters) and forests more patchy and open than other forest types. The richest relicts of boreal rainforests are situated on productive soils (well-drained marine sediments). Most of these forests lie within the southern boreal subzone at under 200 meters elevation.

Particularly well-developed boreal rainforests are found at 63 to ~ 65°N latitude, along the coast of Trøndelag and Helgeland (central Norway, ~63°N latitude), including the valley of Namdalen (see plate 8). These rainforests are dominated by Norway spruce (*P. abies*) mixed with boreal deciduous trees like grey alder (*Alnus incana*), downy birch (*Betula pubescens*), aspen (*Populus tremula*), goat willow (*Salix caprea*), and rowan (*Sorbus aucuparia*). Such forests represent a northern extension of similar forests in Scotland and Ireland.

Lichen communities in these forests have high conservation value because they include several "red-listed" (threatened species as compiled by the IUCN-World Conservation Union) species (Tønsberg et al. 1996; Storaunet et al. 2005; Timdal et al. 2006). A very interesting lichen-flora element termed the "Trøndelag phytogeographical element" corresponds to the lichen rainforest communities of Alaska and British Columbia and the coast of Newfoundland (Holien and Tønsberg 1996; Printzen and Tønsberg 1999).

Most of Norway's boreal rainforests were altered by selective logging in the past, but most forest regenerated naturally until 1950, when plantations of Norway spruce became more widespread. Around 300 relict rainforest patches remain in northern regions, ranging from 2 to 250 hectares,[1] a small fraction of which (less than 1 percent) is protected as nature reserves as part of Norway's forest plan (Håpnes 2003; panda.org[2]), which will be revised soon. In unprotected rainforest with old-growth characteristics, forest owners are in most cases not allowed to use traditional clear-cut logging practices. The forest authorities generally approve plans involving selective felling and very small clearings (less than 0.5 hectare) that attempt to mimic the natural gaps in the forest canopy resulting from tree death. In ravine areas, the possibilities for timber harvest are even more limited. Forest owners do get some economic compensation for leaving trees, though. However, the Norwegian government should protect all remaining rainforest relicts as part of the next phase of forest planning, due out in 2010.

---

[1] www.assets.panda.org/downloads/forestheritage.pdf

[2] www.assets.panda.org/downloads/forestheritage.pdf

Boreonemoral Rainforests

Moderate precipitation, combined with mild climate of Norway's coastline, restricts these rainforests to coastal southwest counties (58–62°N latitude, see figure 6-2), including Rogaland, Hordaland, Sogn og Fjordane, and Möre og Romsdal. Boreonemoral forests are represented by alder (*A. glutinosa*), ash (*Fraxinus excelsior*), oak (*Quercus* spp.), small-leaved lime (*Tilia cordata*), and wych elm (*Ulmus glabra*). Because they include overlapping temperate and boreal forests, the boreonemoral rainforest contains more plant species than any vegetation zone in Norway, making this one of the most biologically diverse regions in all of Fennoscandia (Moen 1999). The epiphytic lichen flora of the boreonemoral rainforests is even richer in species number with affinities that can be traced to the British Isles and the Atlantic Islands (Madeira, Azores). Many species of mosses and lichens reach their global northern limit here (Jørgensen 1996).

No similar mapping or area estimate is available for boreonemoral forests, which are more scattered than their northern counterparts. Examples of boreonemoral rainforest with some "naturalness" (i.e., a semi-natural condition) in terms of trace levels of previous logging can be found in Norway's National Nature Reserves (Storaunet et al. 2005). These rainforest relicts are central to conservation strategies that include both protecting remaining rainforests and restoring their features. For instance, in the new guidelines recently adopted for conservation of Norway's biodiversity by the Norwegian Parliament (Stortinget), restoration of several types of forest is a stated objective toward achieving the overall goal of stopping the loss of biodiversity.

There is no official conservation plan for boreonemoral rainforests at the moment. Few localities have been protected as broadleaf deciduous forest reserves. However, a new evaluation of the Norwegian forest protection plan is forthcoming. Most certainly this will lead to a national mapping of boreonemoral rainforest in Norway and identification of the most scientifically valuable areas for protection.

THE REPUBLIC OF IRELAND AND NORTHERN IRELAND (UK)

Due to the moderating influence of the Gulf Stream, portions of Ireland receive enough rainfall (greater than 2,000 millimeters annually—Cross 2006) to provide suitable conditions for relict rainforest vegetation. On average, rainfall occurs on 190 to 250 days of the year along the east and west coasts of Ireland, respectively, exceeding evapotranspiration for all but one month (Cross 2006).

Because most of Ireland's forests were destroyed centuries ago, describing what these forests may have been like is an imprecise science. Based on peatland deposits and paleo-botanical reconstructions, Ireland's contemporary lowland woodlands generally can be traced back some 8,000 years ago, when they were a mixture of elm and oak with yew (*Taxus baccata*), rowan or ash, and hazel (*Corylus avellana*) in the understory (Ingrouille 1995; Tuite and Brown 1998; Cross 2006). Other woodland species included downy and weeping birch (*B. pendula*), willows (*Salix* spp. on moist sites), alder, English holly (*Ilex aquifolium*), Scots pine (*Pinus sylvestris*), and numerous ferns (Ingrouille 1995; Tuite and Brown 1998; Cross 2006). Periodic climatic changes shifted the species mix over time but the biggest impact to forests came from Neolithic cultures and the onset of the Bronze Age, which offered only brief respites during which forests were allowed to recover (Cross 2006; see box 6-1). During medieval times, Ireland appears to have been dominated largely by scrubby woodland and pastureland mixed with local dense woodlands.

Recent potential vegetation mapping of Ireland may shed some light on forest types and possible locations more closely matching the criteria of rainforests used in this book. Potential vegetation includes the vegetation most likely to be present on a given site based on current climatic conditions in the absence of human intervention. Using this approach, for instance, Cross (2006) identified nine potential vegetation units that may have supported forests before widespread human intervention. Of these, Unit 3—"sessile oak forests rich in bryophytes and lichens"—most closely matches the conditions required for rainforests as described in this book. Therefore, sessile oak forests will be discussed herein as representing the majority of the rainforest relicts of Ireland and having broader importance throughout much of Europe (see box 6-2). For this discussion it should be noted that most of relict rainforests are scattered along the southern and western coastlines of the Republic of Ireland, with a few fragments in Northern Ireland, as noted by Cross (2006).

Sessile Oak Woodlands

According to Cross (2006), the oak woodlands of Ireland are dominated principally by sessile oak (*Q. petraea*), which can grow to 25 meters high on productive sites, with birch (*B. pubescens*) and rowan occupying a canopy position, and alder occurring on wetter sites. A well-developed understory consists of holly, heather (*Calluna vulgaris*), and whortleberry (*Vaccinium mytrillus*), mixed with saplings of birch and rowan. Grasses, forbs, and ferns (*Dryopteris* spp.) are especially prominent, with vines distributed both along the ground and extending into the forest canopy. The rich community of bryophytes and lichens, however,

---

BOX 6-2
Oak Woodlands as High-Conservation-Value Forests of Europe.

Oak woodlands are recognized throughout Europe for their conservation value principally because they contain over 500 species of plants and animals, are rich in bryophytes and lichens, and are regarded as conservation priorities in the United Kingdom Biodiversity Plan.[1] Restricted to very damp, humid areas with high rainfall and acidic soils, they cover some 70,000 hectares in England, particularly the Furness Fells of Cumbria and the southwest; in Scotland, on rocky, exposed western coastlines and south-facing slopes of highland glens; in Wales, along the western coastline where the climate is mild and wet; and in Ireland, where they are widely scattered mainly along the west and southern coast. We did not include oak woodlands in France and Spain, as these forests are generally too dry to meet rainforest definitions used in this book. Oakwoods are rich in neotropical migratory birds, provide nesting sites for buzzards (*Buteo buteo*) and red kites (*Milvus milvus*), food for red deer (*Cervus elaphus*) and roe deer (*Capreolus capreolus*), and habitat for European badger (*Meles meles*), among other species. Centuries of coppicing, cutting for fuel and smelting, and conversion to conifer plantations have nearly eliminated all seminatural oak woodlands. Remaining areas are a priority for restoration, primarily by removing exotics, reducing livestock grazing, and restocking with native species, as recognized by the European Union's Habitats Directive and the Forestry Commission.

---

[1] www.forestry.gov.uk/forestry/Uplandoakwood/

---

is perhaps the most definitive feature of these rainforests, linking them to temperate rainforests globally.

Sessile oak woodlands occur primarily in areas with rainfall over 1,200 millimeters, rainy days in excess of 250 per year, humidity consistently high, temperatures cool in the coldest (6°C) and warmest (15°C) months, and frosts slight and infrequent (Cross 2006). Such conditions, and the sessile oak woodlands they support, are generally found in the extreme southwest of the Republic of Ireland, with the most extensive stands in the mountainous regions of Cork and Kerry counties (51–54°N latitude, Cross 2006). Additional localized fragments can be found along the western coastline in the mountains of Connemara, Donegal, Mayo, and Sligo, on lake islands in blanket bogs, locally on high ground, and in sheltered, humid valleys east (Cross 2006). Other examples include Derrycunihy and Tomies Woods (Killarney National Park, 52°N latitude) and Uragh Wood Nature Reserve (southwest, ~54°N), Glengarriff Woods Nature Reserve (south, ~54°N latitude), Derryclare Wood Nature Reserve (central, ~53°N latitude), Brackloon Wood (northwest, ~53°N latitude), Glenveagh Woods and Glenveagh National Park (central, ~52°N latitude), and Largalinny (north, ~54°N latitude), including Correl Glen National Nature Reserve (~55°N latitude—Kelly and Moore 1975; Kelly 1981; Cross 2006).

Notably, the Killarney region retains the most-extensive (~1,200 hectares) native forest in all of Ireland (Kelly 1981), and Killarney National Park (10,236 hectares) specifically is reported to be within the temperate rainforest zone of Ireland's southwest coastline (Kellogg 1992). The Killarney region also includes an oceanic element of the cryptogamic flora (moss, algae, ferns, fungi). The oceanic element includes the *Blechno-Quercetum scapanietosum* community type (i.e., oak woodland rich in bryophytes) in the wettest areas (Kelly 1981). A significant tract of this forest remains around the Lakes of Killarney within the Park; however, this area was felled around 1800, replanted with oaks, and then replaced with conifer plantations during the twentieth century (Kelly 1981). Additionally, an isolated pocket (25 hectares) of nearly pure yew forest occurs in Reenadinna Wood (~53°N latitude) in the center of the Park, representing the only such forest in all of Ireland. Some of these trees are older than 200 years but grow to a maximum of only 14 meters, with the largest trees reaching a diameter of 125 centimeters (Kelly 1981). The park also contains Ireland's only herd of red deer, a survivor of the last ice age and currently numbering about 850 animals.[3]

The earliest settlers of Ireland arrived some 9,000 years ago and were greeted by oak, hazel, and elm woodlands (Ingrouille 1995; Rackham 1995). Unfortunately, Ireland would undergo an extensive period of deforestation dating back some 6,000 years, during which Neolithic ("Stone Age") cultures first cleared forests for agriculture (see box 6-1; Higgins et al. 2004). Early deforestation was accompanied by a changing climate that selected bogs over woodlands (Tuite and Brown 1998). Aided by metal tools produced during the "Bronze Age" some 2–4,000 years ago, an expanding Irish population cleared nearly all its lowland forests (Higgins et al. 2004). From the eighth to the eleventh centuries, invading Norsemen as well as Cistercian monks (a Roman Catholic clerical order) cleared forests for year-round grazing and agriculture and

---

[3]www.homepage.eircom.net/~knp/; www.pbs.org/wnet/nature/ireland/map.html

opened up trade in the export of oak, which continued for centuries. A policy of deforestation was later adopted by Queen Elizabeth I of England who, during the English Tudor dynasty (1485–1603), , sanctioned the systematic destruction of woodlands in order to gain control over territory (Tuite and Brown 1998).

Within two centuries, Ireland's forest cover had declined from an estimated 12.5 percent to less than 1 percent, making this one of the most heavily deforested regions of Europe (Shepherd 1998; Tuite and Brown 1998; Williams 2003; Higgins et al. 2004). Today, Ireland's surviving semi-natural woodlands, totaling some 80,000 hectares or 1 percent of the forest base, are hanging on by a slim thread (Higgins et al. 2004), although other estimates report nearly 2 percent of the land base in semi-natural woodland (Perrin et al. 2008; also see table 1-2 for estimates generated by the rainforest distribution model). Most of these woodlands are isolated parcels of less than 5 hectares; woodlands more than 20 hectares are exceptionally rare (5 percent) (Higgins et al. 2004). But perhaps the greatest threats to remaining oak woodlands are overgrazing by domestic livestock and the introduction of exotic species, especially common rhododendron (*Rhododendron ponticum*), which shades out native understory plants (Kelly 2005). Unmanaged herds of Sika deer (*Cervus nippon*) overgraze understory plants. Additional threats include air pollution, which is particularly damaging to lichen and bryophyte communities.

While Ireland is experiencing a period of afforestation due to extensive tree planting begun in the 1950s and 1960s, most of this involves commercially grown and exotic conifers such as Sitka spruce (*Picea sitchensis*), lodgepole pine (*Pinus contorta*), and Norway spruce; Scots pine also is commercially grown (Tuite and Brown 1998; Shephard 1998; Higgins et al. 2004; Forestry Commission 2007; DEFRA 2007; see box 6-3). Beginning in the 1980s and leading up to Ireland's ratification of the Convention on Biological Diversity, Coillte (the Republic of Ireland's state forestry body) and the Forest Service of Northern Ireland have been planting selected areas increasingly with broadleaf trees to reestablish semi-natural conditions (Higgins et al. 2004).

Restoring Ireland's forests is a multi-step process aided by recent vegetation inventories designed to locate semi-natural woodland (Higgins et al. 2004) and by reconstruction efforts assisted by potential vegetation mapping to determine local site conditions likely to support restored rainforests (Cross 2006). Equally important is the protection of semi-natural woodlands, as these areas generally have relatively high levels of biological diversity and represent the best hope for restoring more species-rich woodlands.

Beginning in 2001, a Native Woodlands Scheme was adopted by Northern Ireland's Forest Service that includes guidelines for conservation and wood

---

BOX 6-3
European Tree Plantations and Forestry.

Modern plantation forestry has its origins in the early 1600s, when it first appeared in the Alps and then spread across Europe, including Britain and Ireland by 1800 (Ingrouille 1995). European forestry methods, including plantation forestry, spread throughout Europe and were later adopted in the United States in the early part of the twentieth century. Plantations, in general, lack the complexity and richness of native forests, as exotic trees, mostly a few fast-growing conifers such as Douglas-fir (*Pseudotsuga menziesii*), Sitka spruce (*P. sitchensis*), and lodgepole pine imported from western North America, and the hybrid larch (*Larix* x *eurolepis* Henry), are planted in tight rows resembling cornfields more than biologically rich rainforests (also see chapter 2; plate 9ab). European forestry typically includes the use of herbicides and fertilizers and the intensive cultivation of soils (Ingrouille 1995), which, because of multiple logging rotations and the nutrient demands of densely stocked tree farms, have shown signs of diminished productivity in places (Maser 1988). Plantations are managed on very short rotations in order to maximize wood fiber, and they lack the complex features (such as large living and dead trees) and ecological properties of unlogged forests (www.birdweb.net/forestry.html/). Traditional European forestry removes competing vegetation and dead or dying trees, creating a "sterile" appearance, replacing biological complexity with oversimplification of forest structure, process, and function.

---

production (Little et al. 2008). The primary objectives of this program are to conserve existing native woodlands and to create new ones through the restoration of trees and shrubs suitable to the prevailing site conditions; production of quality wood is a secondary objective. Similar actions are taking place throughout Europe through a series of woodland initiatives aimed at sustainable forestry practices (see below).

GREAT BRITAIN

Great Britain has a mild climate owing to the influence of the Atlantic current. Despite this, there is a strong gradient of temperatures leading to a cool summer climate in Scotland that some consider a boreal rainforest climate. However, we considered the rainforests of Great Britain to be perhumid since mean annual temperatures were above those in boreal regions (see figure 1-3 in chapter 1). In particular, the region's oak-woodlands qualify as rainforest communities and are found mainly along the western Atlantic seaboard, the west coast of Scotland, parts of Wales, and possibly southwest England (see figure 6-1). They have many similarities to Ireland's rainforest woodlands (Cross 2006); however, in general, Britain's forests are part of the Celtic broadleaf forest,[4] a region considered critically endangered due to extensive deforestation. A long history (50,000 years or more) of glaciation and the insularity of the region's islands have resulted in low tree and shrub diversity (Ingourille 1995; see box 6-1). The present woodlands are certainly replacement communities of the original pristine post-glacial rainforests. Moreover, past woodlands do not necessarily represent a clear analogue for what should be planted or restored in the future, since there has been marked climate change even in post-glacial periods (Tipping 2003). However, the tree-species composition of the few remaining semi-natural woodlands still reflect some of the major potential vegetation in Britain, particularly those woodlands with understories carpeted by ferns, mosses, and lichens. As in Ireland, zonal rainforests are dominated by oaks (*Q. robur*, *Q. petraea*; see box 6-2) on neutral to acid soils, intermingled with downy and weeping birch (*B. pubescens* and *B. pendula*). European ash (*F. excelsior*) dominates the more base-rich soils, often in association with alluvial forests populated by alder.

Other unique rainforest relicts that have been recently recognized by scientists include hazel woodlands rich in lichens and epiphytes, which are concentrated mainly along the western coastline of Scotland (Coppins and Coppins 2002). In the Scottish Highlands, Scots pine constitutes another native rainforest relict that distinguishes these rainforests from oak types. Also known as "Caledonian Forest," these forests were named because of their ancient Roman name for the wooded heights that once carpeted the Scottish Highlands[5] (Ingourille 1995). Such forests once supported European bison (*Bison bonasus*), wild boar (*Sus scrofa*), lynx (*Lynx lynx*), moose (*Alces alces*—in Europe, moose are called "elk"), red deer, European brown bear (*Ursus arctos arctos*), and European wolf (*Canis lupus lupus*). With the exception of deer, all large mammals were extirpated centuries ago.

Less than 1 percent of the original Caledonia forest remains in scattered, isolated fragments in northwest Scotland (treesforlife.org.uk[6]; Ingourille 1995),

---

[4]www.nationalgeographic.com/wildworld/profiles/terrestrial/pa/pa0409.html
[5]www.treesforlife.org.uk/tfl.contents2.html
[6]Ibid.

which makes Scotland the least-forested country of Europe after Ireland (Foot 2003). At the Beinn Eighe National Nature Reserve in Coille na Glas Leitir (~57°N latitude, plate 10), temperate rainforest is restricted to lower elevations (12–300 meters) on the east side of the reserve.[7] Here, ancient trees live to be older than 400 years and are dominated by Scots pine mixed with birch, rowan, holly, and oak. Remarkably, these woodlands have persisted in a relatively unaltered state for over 8,000 years, but only as isolated fragments totaling some 235 hectares. They contain perhaps the best examples of moisture-loving moss and liverwort communities in all of Great Britain.

Additional examples of temperate rainforest with ancient Caledonia pinewood can be found in the glens (U-shaped glacial valleys) at Glen Affric National Nature Reserve (~57°N latitude) southwest of the village of Cannich in the Highland region of Scotland, approximately 24 kilometers west of Loch Ness. The sheltered glens of Rhum National Nature and Biosphere Reserve (~57°N latitude), the largest of the island quartet on the "Small Isles" of Scotland's western coastline, once included birch woodlands that would qualify as temperate rainforest. Unfortunately, the last native forest was cut down in 1796 for firewood.[8] Efforts are underway to restore the area by planting with native trees. Additional threats to semi-natural woodlands in Scotland include air pollution, which affects tree health and lichen diversity in some places. Regeneration of native species is largely impeded by the overgrazing of sheep and deer. While some degree of grazing can be beneficial to biodiversity, as was proven by the traditional use of wood pastures, nevertheless deer numbers remain above the carrying capacity of most woodlands, as deer represent one of the last-remaining profitable agricultural sectors in Britain. Scottish authorities are now encouraging deer management plans in order to limit the effect of browsing.

Nonnative species are invading stands and threaten to eventually replace natural communities. The most common examples are Sitka spruce, rhododendron, sycamore (*Acer pseudoplatanus*) and, with the exception of southern England, where it is considered native, European beech (*Fagus sylvatica*). Whereas the negative effects of sycamore and beech are under debate, there is no doubt that rhododendron severely reduces biodiversity and the regeneration of native species (Peterken 2001). Measures to eradicate rhododendron are laborious, but have been successful at stand levels.

The attitude toward woodland has changed in recent decades. Consider that whereas during much of the twentieth century woodland was mainly

---

[7] www.nnr-scotland.org.uk/managing_detail.asp?NNRId=15

[8] www.snh.org/uk/publications/online/desinatedareas/nnrs/beinneeighe/beinneighe.asp

treated as a source of income for rural areas, nowadays the recreational benefits and importance of these forests are becoming increasingly apparent. The majority of people are in favor of a higher proportion of land to be reforested, preferably with broadleaf trees or mixed stands of native and conifer species (Mather 2003). Propitiously, there is a growing commitment to increase the percentage of native woodland from governmental institutions, nature conservation agencies, and local communities.

An ambitious goal put forth by the Scottish Forestry Strategy (2006) is designed to increase woodlands from ~17 to 25 percent of the surface area by 2050—a remarkable feat if achieved, considering that about half of Scotland consists of naturally wood-free areas on blanket bogs and waterlogged soils (Tipping 2003). Since the 1970s, when fewer than 10 percent of newly planted trees were native, the ratio of native trees has increased to about 50 percent of the forest base (Worrell and MacKenzie 2003). These are indeed very positive trends; however, there is a big difference between planting native trees and restoring anything resembling natural woodlands. The species choice unfortunately often follows either economic reasoning or personal preferences, without any full consideration of ecological communities. New approaches to planting take soils and extant vegetation into consideration using Ecological Site Classification (Pyatt et al. 2001; Pyatt 2003) to guide the appropriate mix of tree species that best constitute natural woodland communities (see Rodwell 1991). Natural regeneration is always preferable to artificial site preparation; however, local seed sources are not always available (Worrell 1992). Apart from increasing the total area of woodlands in Britain through restorative actions, the protection of remaining relicts is perhaps even more important, because they can serve as a blueprint or reference area from which to re-create semi-natural woodlands.

Another recently established measure to support biodiversity is the creation of Forest Habitat Networks, designed to connect isolated woodland blocks and enable fauna and flora to migrate through the landscape (Worrell and MacKenzie 2003).

CENTRAL EUROPE AND THE NORTHWEST BALKANS

In Central Europe and adjacent areas in the Northwest Balkans, woodlands cover up to 50 percent of the landscape. Only a small fraction of them meet climatic conditions of perhumid rainforest and these are found in three regions, from north to south: (1) Bohemian Forest[9] (Šumava/Bayerischer Wald/Böhmerwald—Central Europe); (2) Northeast Limestone Alps, Prealps (Nördliche Kalkalpen—the Eastern Alps north of the Central Eastern Alps, located in the alpine states of Austria and Germany) and Swiss Prealps (Western Alps); and (3) Southeast Alps and Northwest Balkan ranges in Slovenia (Southeast and Central Europe, see figure 6-3).

The distribution of temperate rainforests in Central Europe is strongly connected with the atmospheric circulation system, primarily during the growing season. Cyclones originating from the North Atlantic strike the Bohemian Forest Range and the northern (outermost) mountain ranges of the Alps, while Mediterranean cyclones move northeast and hit the southeast ranges of the Alps and the northern Balkan Mountains in Slovenia. Beside high annual precipitation (from 1,500 to over 2,000 millimeters), temperate rainforests generally occur on northern exposure over water-retaining soils and/or valley slopes with limited insolation (solar radiation received on the ground).

There is clear evidence from pollen profiles that after the last glaciation the spread of Norway spruce and silver fir (*Abies alba*) as well as European beech and other deciduous broadleaf trees originated from refugia situated on the northern and northwestern Balkans (Draxler 1977; Drescher-Schneider 2003; Pini et. al. 2009). In the Swiss Alps, fir spread from refugia in the northern Apennines of Italy, where it has been the dominant tree species (Burga and Perret 1998). Such climatic refugia provided the evolutionary backdrop for contemporary rainforests to originate.

Bohemian Forest

This rainforest region is limited to below 1,500 meters elevation at latitudes of 48° 30′ to 49° 40′. The old weathered mountain range (16,000 square kilometers; 1.6 million hectares), consisting of crystalline and gneiss bedrock, is situated north of the alpine system. This range spans three countries: Germany (the largest part), the Czech Republic, and Austria. Of this, only about 220,199

---

[9]The Bohemian Forest is a low mountain range in Central Europe extending from South Bohemia in the Czech Republic to Austria and to Bavaria in Germany. The mountains form a natural border between the Czech Republic and Germany and Austria. For historical reasons, the Bohemian and German sides of the forest have different names: in Czech, the Bohemian side is called *Šumava* and the Bavarian side *Zadní Bavorský les*, while in German, the Bohemian side is called the *Böhmerwald* (literally, "Bohemian Forest"), and the Bavarian side the *Bayerischer Wald* (literally, "Bavarian Forest"). In Czech, *Šumava* is also used as a name for the entire adjacent region in Bohemia. (From Wikipedia.)

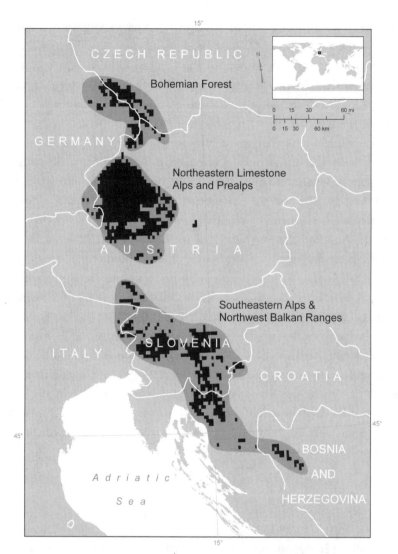

Figure 6-3. Temperate rainforests of three regions of central and southeastern Europe: (1) Bohemian Forest; (2) Northeast Limestone Alps and Prealps (Austria and Germany); (3) Southeast Alps and Northwest Balkan ranges, based on the rainforest distribution model.

hectares provide climatic conditions suitable for rainforests (based on the rainforest distribution model) and these are scattered along slopes of 700 to 1,150 meters elevation. These species-poor rainforests are naturally dominated by European beech, silver fir, and Norway spruce, with lesser amounts of sycamore, European ash, wych elm, lime (*Tilia platyphyllos*), wild cherry (*Prunus avium*), rowan, and yew, which is generally quite rare. Valleys up to 700 meters elevation are covered by spruce, with scattered downy birch. Broad buckler fern (*Dry-*

opteris dilatata) is dominant in the understory; on some stands other ferns (*Dryopteris carthusiana*, *Gymnocarpium dryopteris*, and/or *Athyrium filix-femina*) are present. Originally of less importance, spruce is now the dominant tree in many of these rainforests because of intensive management. Above 1,150 meters, climatic conditions become unfavourable for most rainforest species except for spruce, which is naturally occurring, along with lower amounts of rowan and sycamore (Petermann and Seibert 1979).

Since 1970, 13,000 hectares of these forests have been designated as a National Park in Germany (Bavarian Forest National Park; in 1963 Czechoslovakia (now the Czech Republic) designated a Protected Landscape Area of 97,970 hectares; the most valuable part, with an area of 69,030 hectares, was declared as a National Park in 1991. Only 666 hectares of the park, however, contains virgin forest on the southeast-exposed slope of Mt. Boubin, with a mosaic-gap structure present since 1856 (Leibundgut 1982; Wikipedia[10]). In these reserves, spruce and fir tower above 50 meters. In Austria only two small private reserves totalling 163 hectares exist, and thus more protected areas are needed.

Although the woodlands of the Bohemian Forest have been used intensively for glass production since the Middle Ages, the forests have been increasingly recovering. Today, beside the population of Ural owl (*Strix uralensis*), a stable population of lynx exists in the National Park Area.

Northern Limestone Alps and Swiss Prealps

Meso-climatically suitable sites of the montane altitudinal belt support approximately 745,915 hectares (based on rainforest distribution model) of temperate rainforests in the northern Alps (see plate 11). The scattered forest patches in the Swiss Prealps (Western Alps) are found at approximately 47°N latitude, with the main distribution around the central Swiss Lake District surrounding Luzern. Only 2 percent of the area between 600 and 1,400 meters elevation remains as woodland, originally dominated by silver fir. Forest patches are extremely scattered, separated by extensive pastures. Because of intensive management by the local population, Norway spruce is co-dominant. Harvesting of forest litter, used in farming until the 1960s, and single-stem harvest by private owners has eliminated all but about 6 percent of the woodland area considered in old-growth condition.[11] Although in Switzerland there are forest reserves of 31,301 hectares, this represents only 2.5 percent of the total Swiss forest cover

---

[10]www.en.wikipedia.org/wiki/_umava_National_Park
[11]www.lfi.ch/resultate/schweiz.php

and there is an urgent need for larger patches of highly productive woodland to be represented in protected areas.[12] High population density (~135 people per square kilometer), combined with intensive land management, triggered the demise of lynx and bear in the region.

The eastern Prealps (Eastern Alps), with comparable yearly precipitation of 1,500 to over 2,000 millimeters, and 650 to over 900 millimeters during the growing period,[13] includes still greater amounts of beech in the natural vegetation. The mesic condition of these forests is perhaps best represented by the presence of oceanic lichens such as dot lichen (*Arthonia leucopellaea*—found on conifers) and barnacle lichen (*Thelotrema lepadinum*), which occur along the northern slopes of the Eastern Limestone Alps as well as in few localities in the Swiss Lake District and the Prealps of Fribourg. Moist conditions are maintained by a lack of dry periods during the summer, continuous high air humidity, and a high frequency of fog (Zukrigl et. al. 1963; Schauer 1965). Such conditions, along with the presence of older trees, which act as lichen hosts, allow rainforest lichens to persist. Two other lichens, lung moss (*Lobaria amplissima*) and wart lichen (*Pyrenula laevigata*), have an extremely restricted distribution in the northern Alps (Schauer 1964, 1965).

The greatest portion of temperate rainforest in this region occurs in the provinces of Salzburg and Upper Austria; however, these forests have been exploited through the mining of metals and salt since the Bronze Age. In a remote part of Lower Austria, a small (275 hectares) virgin forest ("Rothwald"—47° 47′ N latitude) south-southeast of Lunz (west of Mariazell) has been protected since 1875 as wilderness. This woodland lies on limestone and is situated between 950 and 1,500 meters elevation. It also has been included in the NATURA 2000 network ("Ötscher-Dürrenstein") since 1998. Gap dynamics and changes in tree-species composition in the course of stand renewal have been studied for decades in these forests (Zukrigl et al. 1963; Neumann 1978; Mayer et al. 1980).

A variant of montane mixed woodlands also occurs in this region and is noteworthy for its tall and mesic forb and fern understories. The mixed woodland consists of spruce, fir, and beech, with Austrian spurge (*Euphorbia austriaca*, endemic of the Northeast Alps), columbine meadow rue (*Thalictrum aquilegifolium*), monkshood (*Aconitum* spp.), round-leaved saxifrage (*Saxifraga rotundifolia*), alpine bells (*Cortusa matthioli*), alpine blue-sow thistle (*Cicerbita alpine*), and

---

[12]www.bafu.admin.ch/umwelt/daten/04564/index.html?lang=de

[13]www.zamg.ac.at/fix/klima/oe7100/klima2000/klimadaten_oesterreich_1971_frame1.htm

musk thistle (*Carduus personata*). Canopy gaps support species such as perennial honesty (*Lunaria rediviva*), Hart's-tongue fern (*Asplenium scolopendrium*), and Austrian caraway (Österreichkümmel, *Pleurospermum austriacum*). Wildlife include brown bear, lynx, chamois (*Rupicapra rupicapra*), red deer, golden eagle (*Aquila chrysaetos*), rock ptarmigan (*Lagopus muta*), capercailzie (*Tetrao urogallus*), and black grouse (*Lyrurus tetrix*) (Kraus 1991; Aste and Gossow 1996; Rauer and Gutleb 1997).

Southeastern Alps and the Northwest Balkans

The Southeastern Alps (Carinthia/Austria, Republic of Slovenia) and the karst of the Northwestern Balkans (Republic of Slovenia, northern Croatia—45° 30′–46° 30′ N latitude) support about 577,425 hectares (based on the rainforest distribution model) of temperate woodlands dominated by beech, extending to timberline around 1,550 meters elevation (Snežnik in Slovenia and Risnjak in Croatia are the highest peaks of the northern Dinaric Alps). Although strongly influenced by Mediterranean climate (two maxima of precipitation in spring and autumn), high annual rainfall precludes drought. Taxa of Illyrian[14] distribution (mainly *Anemone trifolia*, *Hacquetia epipactis*, *Omphalodes verna*, *Epimedium alpinum*, *Lamium orvala*) are characteristic of these species-rich and moist woodlands. Lichens with a Central European–Mediterranean montane and oceanic distribution also occur and include lung lichen (*Lobaria scrobiculata*) and jelly lichen (*Collema fasciculare*). Wildlife include brown bear (~500 individuals), wild boar, and, in the sparsely populated remote mountains, wildcat (*Felis silvestris silvestris*), European wolf, lynx, and Ural owl.

While Slovenian forests underwent great changes in the past 50 years, only 15 percent of the forests (mainly coppiced stands) have been converted to plantation conifers,[15] and more than one-third of the Slovenian territory is protected within the framework of NATURA 2000. (However, see table 10.1 for different estimates using stricter protected-areas criteria.) Notably, nearly 60 percent of this territory is forested, including reforested mountain pastures.

A tract of 60 hectares of virgin forest, "Pečka," occurs in the Kocevje Mountains (Slovenia) between 740 and 940 meters elevation on sites with a high geomorphological differentiation caused by karst weathering. Some

---

[14]Illyrian languages are a group of Indo-European languages that were spoken in the western part of the Balkans. Species of this distribution type range from the southeast Alps to the northwest Balkans.

[15]www.waldwissen.net/themen/wald_gesellschaft/weltforstwirtschaft/wsl_waelder_sloweniens_DE

patches of this generally dense woodland show a higher cover of ferns (*D. filix-mas*, *Asplenium* = *Phyllitis scolopendrium*, *Sanicula europaea*) and other mesic species.

Conservation priorities for these rainforests in general should begin with connecting the few remaining virgin forests with already-protected woodlands to enable the exchange of plant and animal populations, especially as the global climate shifts. In particular, the northwestern Balkan area could serve as a migration corridor to the eastern Alps for bear, wolf, and lynx, all of which were extirpated during the nineteenth century but with colonizing populations nearby or recently reintroduced. Lynx were reintroduced to the eastern Alps, and Ural owl and bear are known to cross the border from Slovenia into Austria. Moreover, migration of lynx from the Bohemian Forest and Slovenian territory was supported by the reintroduction of nine individuals in the 1980s in the "Gurktaler Alpen" region (Central Eastern Alps). An unsuccessful reintroduction of bears, starting in 1989, failed because of lack of support from the regional population (farmers with herds of sheep) and hunters, who illegally shot them. In Switzerland, there is a lynx population of more than 100 individuals in parts of the Alps and in the Jura Mountains (north of the Alps), and since 2005 bears have been observed as a result of migration from northern Italy to the Grisons (Graubünden, the largest and easternmost canton of Switzerland adjoining Italy, Austria, and Liechtenstein). Wolves also have been dispersing from Italy into Central Europe (Hutchinson 2001).

While Germany and the Czech Republic share protected areas in the Bohemian Forest and protection levels seem satisfactory in Austria, there is an urgent need for larger protected areas comprising all forest types and connecting the three regions. Especially important to the region's forests is the protection of old trees and deadwood, which is necessary to conserve rainforest lichens and old-forest species like the black woodpecker (*Dryocopus martius*).

CLIMATE CHANGE

Rainforest relicts, like temperate rainforests throughout the world, are particularly vulnerable to changes in global climate that may disrupt the seasonal distribution of rainfall and ideal climatic conditions that have given rise to these rainforests millennia ago. Recent climate-simulation modeling, for instance, predicts that the wet regions of Scotland and Wales will experience drier summers, milder and wetter winters, more-severe wind storms (although this is more difficult to predict with certainty), and flooding events (Ray 2008a,b).

a

b

Plate 1. Contrasting views of coastal temperate rainforest on the Tongass National Forest, Alaska: (a) intact forest on Admiralty Island, Alaska, and (b) fragmented forest on northeast Chichagof Island. Source: John Schoen.

Plate 2a. One of the world's last remaining relatively intact temperate rainforests, the Great Bear Rainforest, British Columbia, is globally significant. Source: Tim Greyhavens.

Plate 2b. Kermode or "Spirit bear" (*Ursus americanus kermodei*) from Gribbell Island, north coast of British Columbia, considered one of four important nearshore islands on the coast for this unique subspecies. Source: Tim Greyhavens.

Plate 3. Temperate and boreal rainforests are rich in lichens like this witch's hair (*Alectoria sarmentosa*), photographed from the Wind River Canopy Crane Research Facility near Carson, Washington. Source: Dominick DellaSala.

Plate 4a. An ancient western red cedar (*Thuja plicata*) in the Incomappleux River of inland British Columbia, estimated at over 1,800 years old and with a diameter at breast height of 326 centimeters. Intact old-growth rainforests once dominated inland British Columbia but are now increasingly rare as these forests are logged. Source: Craig Pettitt.

Plate 4b. Mountain caribou (*Rangifer tarandus*) swimming across Quesnel Lake, inland British Columbia, considered the deepest glacial-fjord lake in the world. Caribou feed on lichens found primarily in older forests and venture down into valley areas to find forage in the spring and fall. Source: Elysia Resort, British Columbia.

Plate 5a. Humid boreal rainforest like this one in Fitzgerald's Pond Provincial Park, near Argentia, Newfoundland, are dominated by balsam fir (*Abies balsamea*) and black spruce (*Picea mariana*), with understories of mosses (*Sphagnum* spp., *Pleurozium schreberi, Hylocomium splendens*), sometimes with ferns (*Osmunda cinnamomea*), and numerous oceanic lichens. Despite their small stature, such forests have a distinctive biota and are in need of concerted conservation efforts. Source: John McCarthy.

Plate 5b. Lichen community (*Erioderma mollissimum*, in the center; also *Lobaria pulmonaria, L. quercizans, L. scrobiculata, Pannaria rubiginosa*, and *Pseudocyphellaria perpetua*) on red maple (*Acer rubrum*) in a wet/hemiboreal rainforest of Eastern Canada in Shelburne County, Nova Scotia. Source: Robert P. Cameron.

Plate 6. Mixed forest of Coigüe (*Nothofagus dombeyi*) and *alerce* (*Fitzroya cupressoides*) at Nahuel Huapi National Park, Rio Negro Province, Argentina. *Alerce* are capable of living to over 3,500 years, rivaling California redwoods in stature and age. Ancient forests like this one are in decline regionally and globally from logging. Source: Daniel Gomez.

Plate 7. Monkey-puzzle tree (*Araucaria araucana*) in Conguillio National Park, Chile. Old enough to outlast the coming and going of dinosaurs and the splitting apart of ancient Gondwana, this species is considered a "living fossil." Source: Paul Alaback.

Plate 8. Boreal rainforest from the Homla River Valley in Malvik municipality of Sor-Trondelag County, Norway, near the southern limits of this rainforest type. Previously logged areas (30 years old) are seen here as deciduous trees (light green) mixing with old-growth spruce forests, a rarity throughout Europe. Source: Jon Arne Saeter.

a

b

Plate 9. Landscape (a) and stand level (b) views of biologically simplistic and impoverished tree plantations, photographed on the Isle of Skye along the central west coast of Scotland, 160 km west of Inverness and near Glen Affric, 35 kilometers west of Inverness in north-central Scotland, respectively. Source: Paul Alaback.

Plate 10. Beinn Eighe National Nature Reserve on the west coast of Scotland includes outstanding examples of old Scots pine (*Pinus sylvestris*) – downy birch (*Betula pubescens*) rainforests. Source: Paul Alaback.

Plate 11. Probable rare, virgin woodland of Norway spruce (*Picea abies*), European beech (*Fagus sylvatica*), and European larch (*Larix deciduas*), with lung lichen (*Lobaria pulmonaria*) growing on Sycamore maple (*Acer pseudoplatanus*) (right foreground) at the northern slopes of Mt. Zinödl (Gesäuse) of the Northern Limestone Alps, Austria. Source: Anton Drescher.

a

b

Plate 12. A representative cool-temperate rainforest and large Japanese beech (*Fagus crenata*) (a) from the northeastern portion of Honshu, the largest island of Japan (photo credit, Yukito Nakumara). Japanese temperate rainforests contain numerous endemics such as this *Rhododendron albrechtii* (b) near Mt. Naeba, along the border of Nagano and Niigata prefectures in central Honshu, Japan. Source: Yukito Nakumara.

a

b

Plate 13. Two views of mixed temperate rainforest dominated by Australian oak (*Eucalyptus obliqua*) near the Picton River (a) and the Tahune Forest Reserve (b), both in Tasmania. Mixed rainforests like this include some of the tallest hardwoods and most carbon-dense forests in the world, yet they are being logged and exported as wood chips. Source: James Kirkpatrick (a), Tim Greyhavens (b).

Plate 14. Colchic rainforest with beech (*Fagus orientalis*) and *Rhododendron ponticum* in Adjara, Georgia. The region is considered a "hot spot" of biological diversity due to high numbers of plants with ancient affinities and high levels of endemism. Source: Zurab Nakhutsrishvili.

Plate 15. Mixed broadleaf Korean pine (*Pinus koraiensis*) forests (a) on the slopes of Sikhote-Alin Mountains, Russian Far East; and subalpine Siberian pine (*Pinus sibirica*) and Siberian fir (*Abies sibirica*) forests (b) at higher elevations of Sayany and Altai Mountains, Inland Southern Siberia. Source: Pavel Krestov.

Plate 16. Large yellowwood (*Podocarpus latifolius*) over 50 meters tall in the Knysna–Tsit-sikamma forests of South Africa. Poorly protected and small in spatial extent, these forests are highly threatened. Source: Paul Hosten.

Such changes are likely to trigger widespread shifts in tree-species composition, outbreaks of insect and pest species, wildland fire, and loss of epiphytes through desiccation, wind storms, and tree fall in places.

This uncertain future presents a call to action for stepping up conservation measures while there are still viable options. As such, restoring the underlying properties and functions of semi-natural woodlands throughout Europe, which allow relict rainforests to be both resilient and resistant to climate change, is of utmost importance (additional recommendations are covered in chapters 10 and 11). This includes restoring floodplain connectivity to accommodate flooding events from more intense storms anticipated in places, and reconnecting woodland landscapes to enable climate-related forced migrations of wildlife. Scotland's noteworthy goal to expand woodlands from 17 to 25 percent by 2050 is perhaps a model for restoring semi-natural woodlands with some degree of landscape and riparian connectivity. Moreover, Ireland's woodland scheme should play a role in regional cap-and-trade policies through biological sequestration and long-term storage of carbon in forests.

RECONSTRUCTING EUROPEAN RAINFORESTS— "BACK TO THE FUTURE"

Nearly wiped from the face of the Earth by centuries of human exploitation, Europe's few remaining relict rainforests are highly endangered (see box 6-4). European forests once provided the masts for ancient ships, charcoal for iron smelting to make tools and weaponry, hunting grounds for royalty and common people, firewood and timbers for homes, barter for early timber traders, fuel for the industrial revolution, and tannin (stripped from tree bark) for leather processing (Williams 2003). They were cleared for political conquest, sheep and cattle grazing, agriculture, and simply because they got in the way of expanding civilizations and conquests. Despite this inauspicious past, there are encouraging signs ahead. In Scotland and Ireland, trees are being planted in large numbers and areas surrounding native woodlands are being fenced off to enable natural regeneration.

However, even though reforestation is now a stated policy of European nations (Shephard 1998; DEFRA 2007; Forestry Commission 2007), it needs to be based on the requirements of native plant communities and followed by the reintroduction of keystone species like beaver (*Castor canadensis*), which were hunted to near-extinction throughout Europe but are essential to fully functional riparian areas. Pilot work by Scotland's environmental ministry to

---

BOX 6-4

Threats to European Rainforest Relics: Going, Going, Gone?

- Most rainforest relicts were eliminated centuries ago by early settlement, deforestation, political and religious conquest, and intense grazing.
- Nitrogen deposition from atmospheric pollution threatens lichenrich areas throughout Europe (Ingrouille 1995).
- Conversion of rainforest relicts to fast-growing plantations jeopardizes the few remaining semi-natural areas, although this is slowing due to growing interest in sustainable forestry and restoration of semi-natural woodlands.
- Intense livestock grazing, high stocking densities of deer, and invasive species threaten native plant communities.
- Poaching of wildlife, combined with the loss and fragmentation of wildlife habitat, has devastated native populations.
- Few if any areas are intact or strictly protected.

---

reintroduce beavers has recently begun and is an important step in restoring riparian and wetland functions. The reintroduction of other mammals, along with stepped-up public outreach, needs to include lynx, boar, bear, and wolf.

Throughout Europe as well as within particular rainforest relicts, specific conservation measures need to be broadly adopted to protect and restore remaining semi-natural woodlands as part of local biodiversity planning, reforestation initiatives, and scientific research and monitoring. Many ancient woods have been felled and replanted with nonnative commercial conifers, yet remnants of the original forest survive and could be restored before another rotation of conifers is planted.[16] Designating new Natural Heritage Areas and Special Areas of Conservation, and restoring semi-natural woodland in National Parks and National Nature Reserves, among other measures, should be a top priority of national efforts to restore semi-natural woodlands. This should include minimizing grazing and managing the spread of invasive species.

Sadly, Europe's long history of environmental degradation is a stark reminder that we must never let the rest of the world's rainforest get to this point.

---

[16]E.g., see www.woodlandtrust.org.uk/en/why-woods-matter/restoring/PAWS percent20 research/Pages/research.aspx

Scientists and educators need good examples of semi-natural relics, protected from logging and grazing, to learn more about how and why these forests developed in widely scattered corners of the globe and how best to restore them. Places like Ireland illustrate that near-total ecosystem collapse can occur within a single human lifetime, breaking the cultural bond between rainforests and local people. Europe's rainforest relics are therefore "lifelines" to a restored continental rainforest network and a local connection to trees and rainforest communities for all Europeans. These relict forests will prove crucial in developing better strategies to conserve and restore rainforests throughout the world as climate change alters the very conditions that created such rainforests in the first place.

LITERATURE CITED

Alaback, P.B., and J. Pojar. 1997. Vegetation from ridgetop to seashore. Pp. 69–88 in *The rainforests of home*. Ed. P. K. Schoonmaker, B. von Hagen, and E. C. Wolf. Washington, DC: Island Press.

Alfnes, E., and E. J. Førland. 2006. Trends in extreme precipitation and return values in Norway 1900–2004. Norwegian Meteorological Institute. Meteorological Report Number 2/2006. Norway: Blindern.

Aste, C., and W. Gossow. 1996. Habitat suitability for brown bears in the Austrian Alps and potential impairments by forestry and ungulate game conservation. Pp. 284–89 in *Proceedings of the IX International Conference on Bear Research and Management*. Grenoble, France.

Burga, C. A., and R. Perret 1998. Vegetation und Klima in der Schweiz seit dem jüngeren Eiszeitalter. Thun, Switzerland: Ott Verlag.

Coppins, A. M., and B. J. Coppins. 2002. Indices of ecological continuity for woodland epiphytic lichen habitats in the British Isles. British Lichen Society.

Cross, J. R. 2006. The potential natural vegetation of Ireland. *Biology and Environment: Proceedings of the Royal Irish Academy* 106B (2):65–116.

DEFRA [Department for Environment Food and Rural Affairs]. 2007. A strategy for England's trees, woods and forests. www.defra.gov.uk/wildlife-countryside/rddteam/forestry.htm

Draxler I. 1977: Pollenanalytische Untersuchungen von Mooren zur spät- und postglazialen Vegetationsgeschichte im Einzugsgebiet der Traun. *Jahrbuch der Geologischen Bundesanstalt* 120:131–63.

Drescher-Schneider, R. 2003: Die Vegetations- und Klimageschichte der Region Eisenerz auf der Basis pollenanalytischer Untersuchungen im Leopoldsteiner See und in der Eisenerzer Ramsau. Pp. 174–97 in *Archäologische Erforschung der Region Eisenerzer Alpen*. Ed. S. Klemm. Wien: Österreichische Akademie der Wissenschaften, Philosophisch-Historische Klasse, Mitteilungen der Prähistorischen Kommission.

Foot, D. 2003. The twentieth century: Forestry takes off. Pp. 158–94 in *People and woods in Scotland: A history.* Ed. T. C. Smout. Edinburg: Edinburgh Univ. Press.

Forestry Commission. 2007. Choice of Sitka spruce seed origins for use in British forests. Forestry Commission Bulletin 127.

Håpnes, A. 2003. Background note: Natural forest heritage in Norway. WWF Norway.

Higgins, G. T., J. R. Martin, and P. M. Perrin. 2004. National survey of native woodland in Ireland. A report submitted to National Parks and Wildlife Service, Department of the Environment, Heritage and Local Government.

Holien, H., and T. Tønsberg 1996. Boreal rain forest in Norway: The habitat for lichen species belonging to the Trøndelag phytogeographical element. *Blyttia* 54:157–77. (In Norwegian with an English summary).

Hutchinson, J. 2001. Wolves of the world. Wolves in Europe: The action plan for wolf conservation in Europe. *International Wolf* Winter 2001:16–17.

Ingrouille, M. 1995. *Historical ecology of the British flora.* London: Kluwer Academic Publishers.

Jørgensen, P. M. 1996. The oceanic element in the Scandinavian lichen flora revisited. *Symbolae Botanicae Upsaliensis* 31 (3):297–317.

Kellogg, E. 1992. Coastal temperate rainforests: Ecological characteristics, status and distribution worldwide. A working manuscript. Occasional paper series No. 1. Ecotrust, Portland, OR.

Kelly, D. L., and J. J. Moore. 1975. A preliminary sketch of the Irish acidophilous oakwoods. Pp. 375–87 in *Colloques phytosociologiques III: La vegetation des forêts caducifoliees acidophiles.* Ed. J. M. Gehu. Vaduz, Liechtenstein: J. Cramer.

———. 1981. The native forest vegetation of Lillarney, southwest Ireland: An ecological account. *Journal of Ecology* 69:437–72.

———. 2005. Woodland on the western fringe: Irish oak wood diversity and the challenges of conservation. *Botanical Journal of Scotland* 57:21–40.

Kraus E., ed. 1991. Forschungsbericht Braunbär 1. Forschungsinstitut WWF Österreich, Bericht 2/1991. Wien.

Leibundgut, H. 1982. *Europäische Urwälder der Bergstufe.* Bern, Stuttgart: Paul Haupt Verlag.

Little, D., E. Curran, T. Horgan, J. Cross, M. Doyle, and J. Hawe. 2008. Establishment, design and stocking densities of new native woodlands. Native Woodland Scheme Information Note No. 5. Forest Service, Ireland.

Maser, C. 1988. *The redesigned forest.* San Pedro, CA: R. and E. Miles.

Mather, A. 2003. The future. Pp. 214–24 in *People and woods in Scotland: A history.* Ed. T. C. Smout. Edinburg: Edinburgh Univ. Press.

Mayer, H., M. Neumann, and Sommer H. G. 1980: Bestandesaufbau und Verjüngungsdynamik unter dem Einfluß natürlicher Wilddichte im kroatischen Urwaldreservat Corkova Uvala/Plitvicer Seen. *Schweizerische Zeitschrift für Forstwesen* 131:45–70.

Moen, A. 1999. *National atlas of Norway: Vegetation.* Norwegian mapping authority, Hønefoss. ISBN 82-7945-000-9.

Neumann, M. 1978. Waldbauliche Untersuchungen im Urwald Rothwald/Niederösterreich und im Urwald _orkova Uvala/Kroatien. Doctoral thesis. Universität für Bodenkultur, Wien (Univ. of Natural Resources and Applied Life Sciences, Vienna).

Perrin, P., M. Martin, S. B. O'Neill, F. McNutt, and A. Delaney. 2008. National survey of native woodlands 2003–2008. Unpublished report to the National Parks and Wildlife Service, Dublin.

Peterken, G. F. 2001. Ecological effects of introduced tree species in Britain. *Forest Ecology and Management* 141:31–42.

Pini R., C. Ravazzim, and M. Donegana. 2009. Pollen stratigraphy, vegetation, and climate history of the last 215ka in the Azzano Decimo core (plain of Friuli, north-eastern Italy). *Quaternary Science Reviews* 28:1268–90.

Printzen, C., and T. Tønsberg 1999. The lichen genus *Biatora* in northwestern North America. *Bryologist* 102:692–713.

Pyatt, D. G., D. Ray, and J. Fletcher. 2001. *An ecological site classification for forestry in Great Britain*. Bulletin 124. Edinburgh: Forestry Commission.

———. 2003. Interpretation, prediction and modelling with NVC. Pp. 59–86 in *National vegetation classification: Ten years' experience using the woodland section*. Ed. E. Goldberg. JNCC-Report No. 335.

Rackham, O. 1995. Looking for Ancient Woodland in Ireland. Pp. 1–12 in *Woods, trees and forests in Ireland*. Ed. J. R. Pilcher and S. Macantsaoir. Dublin: Royal Irish Academy.

Rauer, G., and B. Gutleb. 1997. Der Braunbär in Österreich. Umweltbundesamt Wien. *Monographien* 88.

Ray, D. 2008a. Impacts of climate change on forestry in Scotland: A synopsis of spatial modeling research. Research Note, Forestry Commission Scotland.

———. 2008b. Impacts of climate change on forestry in Wales. Research Note, Forestry Commission Wales.

Rodwell, J. S., ed. 1991. *British plant communities*. Vol 1.: *Woodlands and scrub*. Cambridge: Cambridge Univ. Press.

Rolstad, J., I. Gjerde, K. O. Storaunet, and E. Rolstad. 2001. Epiphytic lichens in Norwegian coastal spruce forests: Historic logging and present forest structure. *Ecological Applications* 11:421–36.

Schauer, T. 1964. Zur epiphytischen Flechtenvegetation der Umgebung von Lunz (Niederösterreich). *Verhandlungen der zoologisch-botanischen Gesellschaft in Wien* 103:191–200.

———. 1965: Ozeanische Flechten im Nordalpenraum (Doctoral thesis). *Portugaliae Acta Biologica* B 8:17–229.

Scottish Forestry Strategy 2006. SE/2006/155. Laid before the Scottish Parliament by the Scottish Ministers, October 2006. Forestry Commission Scotland ISBN-0 85538 705 X.

Shephard, G. 1998. Europe: An overview. Pp. 3–26 in *The EU tropical forestry sourcebook*. Ed. G. Shepherd, D. Brown, M. Richards, and K. Schreckenberg. London: Overseas Development Institute. www.odi.org.uk/projects/98-99-tropical-forestry/Sourcebook/index.html

Storaunet, K. O., J. Rolstad, I. Gjerde, and V. S. Gundersen. 2005. Historical logging, productivity, and structural characteristics of boreal coniferous forests in Norway. *Silva Fennica* 39:429–42.

Timdal, E., H. Bratli, R. Haugan, H. Holien, and T. Tønsberg 2006. Lichens. Pp. 129–39 in

2006 Norwegian Red List. Ed. J. A. Kålås, Å. Viken, and T. Bakken. Norway: Artsdatabanken.

Tipping, R. 2003. Living in the past: Woods and people in prehistory to 1000 BC. Pp. 14–39 in *People and woods in Scotland: A history.* Ed. T. C. Smout. Edinburg: Edinburgh Univ. Press.

Tønsberg, T., Y. Gauslaa, R. Haugan, H. Holien, and E. Timdal. 1996. The threatened macrolichens of Norway. *Sommerfeltia* 23:258.

Tuite, P., and D. Brown. 1998. Pp. 233–44 in *The EU tropical forestry sourcebook.* Ed. G. Shepherd, D. Brown, M. Richards, and K. Schreckenberg. London: Overseas Development Institute. www.odi.org.uk/projects/98-99-tropical-forestry/Sourcebook/index.html

Williams, M. 2003. *Deforestation of the Earth: From prehistory to global crisis.* Chicago: Univ. of Chicago Press.

Worrell, R. 1992. A comparison between European continental and British provenances of some British native trees: Growth, survival and stem form. *Forestry* 65 (3):253–80.

———, and N. MacKenzie. 2003. The ecological impact of using the woods. Pp. 195–213 in *People and woods in Scotland: A history.* Ed. T. C. Smout. Edinburgh: Edinburg Univ. Press.

Zukrigl, K., G. Eckhard, and J. Nather. 1963. Standortkundliche und waldbauliche Untersuchungen in Urwaldresten der nördlichen Kalkalpen. Mitteilungen der forstlichen. *Bundesversuchsanstalt* 62:1–24, 40–245, and tables.

CHAPTER 7

Temperate Rainforests of Japan

Yukito Nakamura, Dominick A. DellaSala, and Paul Alaback

With a rich human history dating back to a Paleolithic time some 30,000 years ago, Japan is home to an astonishing array of plant communities that range from alpine to subtropical and temperate rainforest (Hämet-Ahti et al. 1974; Box 1995). The Japanese archipelago (~377,853 square kilometers) occupies a fraction (~0.8 percent) of the earth's terrestrial surface and is roughly one twenty-fifth the size of the United States. It is located in a transitional zone between subtropical and subboreal at 30 to 45°N latitude (see figure 7-1). Japan's mountainous island chain rises to 3,000 meters elevation, where forest transitions to alpine tundra. The lack of glacial influence has served as refugia for numerous species, including many endemics that required humid places during the Pleistocene glacial period.

GLOBAL ACCOLADES AND RAINFOREST VITALS

From a global perspective, the temperate rainforests of Japan are exceptional in several ways (see table 7-1). This is the only Northern Hemisphere rainforest where dwarf bamboo species (e.g., *Sasa* spp., *Sasamorpha* spp., *Pleioblastus* spp.) play a key role in the structure and dynamics of both warm and cool-temperate forests, similar to that of the related solid-stemmed bamboos (*Chusquea* spp.) in Chile and Argentina (Nakashizuka 1987). The cool-temperate forests are the only fully developed mesophytic (requiring a moderate amount of moisture)

181

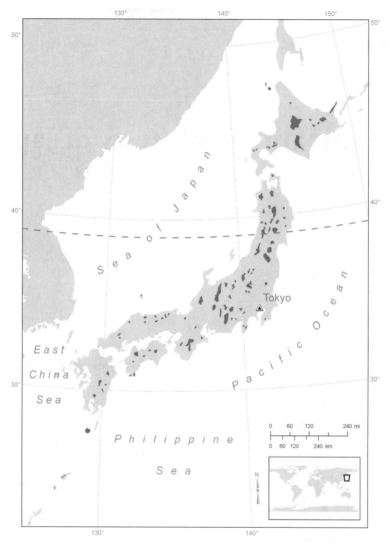

Figure 7-1. Temperate rainforests of Japan (adapted from Miyawaki et al. 1980–1989). Dashed line generally depicts the demarcation between warm-temperate evergreen broadleaf rainforest (south of dashed line) and cool-temperate summer evergreen broadleaf rainforest (north of the line).

deciduous forest that we classify as temperate rainforest (Rohrig and Ulrich 1991). Most of Japan's forest trees, in fact, are closely related to those in the deciduous forests of eastern North America and Europe. Some of the broadleaf evergreen warm-temperate rainforests have a strong subtropical affinity and are the lowest latitude and among the warmest of temperate rainforests described in this book (with a mean temperature during the warmest quarter of 18°C; see

Table 7-1. Global attributes of Japan's temperate rainforests.

Attribute	Importance
Distributed zonally in mountainous islands spanning 3,000 meters elevation along two zones: warm- and cool-temperate	Exceptional beta diversity (species turnover along gradients), plant richness, and endemism
Influenced by oceanic climates delivering abundant (1,200–2,800 millimeters) rainfall, primarily during seasonal monsoons	Allows for persistence of rainforest communities
Unique species for temperate rainforests (e.g., Japanese macaques and Japanese giant salamander)	One of only two temperate rainforests globally with primates; largest aquatic salamander in the world

table 1-1 in chapter 1). Even some of the cool-temperate rainforests are among the warmest in the Northern Hemisphere (with a mean annual temperature of 9°C; see figure 1.3a in chapter 1). Southern and central Japan and parts of the mainland coast are also unique in having dry winters and wet, warm summers, in contrast to most other rainforest regions which have either well-distributed precipitation or have their heaviest rains in fall or winter. This strong seasonal pattern of rainfall also explains the extensive development of deciduous forest in the region.

Approximately 5,565 species of vascular plants are found throughout the Japanese archipelago (United Nations Development Programme et al. 2000), existing at higher levels than in continental temperate regions of comparable size, such as Spain (5,050 species, 504,782 square kilometers) and Germany (2,632 species, 357,021 square kilometers); although Italy has slightly higher species richness for its size (5,599 species, 301,230 kilometers). Perhaps the most striking feature of the Japanese archipelago is the extraordinary number (36 percent) of endemic vascular plants. A long period of isolation from the mainland, combined with periodic volcanic eruptions and complex environmental gradients, have served as an evolutionary "cradle" for speciation events over time. Stable, mild oceanic climatic conditions permitted ancient species to persist or to evolve into distinct species, and rapid diversification has occurred across distinct climatic zones, despite the relatively young geological age of mountains across the archipelago (Yamanaka 1979; Box 1995; Qian and Ricklefs 2000). For example, even though many alpine habitats are quite recent (less than 2 million years old), over half the species in these habitats have evolved into unique (endemic) species since these habitats were formed (Shimizu 1983).

Today, these zones house highly diverse ecosystems, from species-poor snow forests and subalpine conifers to richly diverse broadleaf evergreen subtropical forests. In addition, Japan's warm–temperate rainforests harbor one of the world's only temperate rainforest primates, the Japanese macaque (*Macaca fuscata*). (The Knysna-Tsitsikamma rainforests of South Africa also harbor Chacma baboons [*Papio ursinus*] and vervet monkeys [*Cercopithecus aethiops*]; see chapter 9.) Japanese temperate rainforests are home to the world's largest aquatic salamander, the Japanese giant salamander (*Andrias japonicus*), which can reach nearly 2 meters in length.

ISLAND BIOGEOGRAPHY, GLACIERS, AND VOLCANISM

Oceanic conditions and Ice Age refugia have played dominant roles in the maintenance of the extraordinary diversity of Japanese plants. Volcanic activity, a result of crustal variation, has also influenced dispersal and migration of numerous species across the archipelago. There are many active volcanoes in Japan, the peak of volcanism occurring during the Quaternary (~1.8 million years ago) when the Pacific and Eurasian plates collided. The Japanese archipelago is located at the fringe of the Eurasia plate, which ventured far from the Eurasian continent ~14.5 million years ago, when the continents drifted apart. The Japan Sea provided an isolation barrier that, along with heavy snows in mountainous areas, limited temperate rainforests to Pacific coastal areas of the archipelago and led to the evolution of many unique species.

Like the Alaskan rainforest (see chapter 2), Japan's island insularity influenced the differentiation of subspecies such as goldenrod (*Solidago virgaurea* ssp. *asiatica*), Sika deer (*Cervus nippon* ssp. *yakushimae*), and Japanese macaques (*M. fuscata* ssp. *yakui*). Many endemic plants occur on the Honshu mainland and across the archipelago (see table 7-2), while endemic animals such as the giant salamander occur in warm-temperate areas, dormouse (*Glirulus japonicus*) in cool-temperate areas, and Japanese serow (*Capricornis crispus*) in the dense woodlands of Honshu (see rainforest classifications below).

REGIONAL CLIMATE

The Japanese archipelago is bounded by the Pacific Ocean, the Japan Sea, the Sea of Okhotsk, the East China Sea, and the Philippine Sea. Oceanic climatic conditions are amplified by the warm, tropical Philippine current. The warm and cold currents run into each other at midlatitude, approximately 38°N

Table 7-2. Representative plant species by temperate rainforest zones (warm temperate, cool montane, and subalpine) of Japan. (Asterisks refer to endemics.)

Rainforest Zone	Common Name	Scientific Name
Warm Temperate Evergreen Broadleaf	Japanese chinquapin	*Castanopsis cuspidate & C. sieboldii*
	evergreen oak	*Q. acuta, Q. sessilifolia, Q. myrsinaefolia, Q. glauca, Q. salicifolia,* and *Q. gilva*
	Japanese false oak	*Lithocarpus edulis* and *L. glabra*
	Japanese machilus	*Persea thunbergii*
	Japanese silver tree	*Neolitsea sericea*
	wild cinnamon	*Cinnamomum japonicum*
	Japanese ternstroemia	*Ternstroemia gymnanthera*
	Japanese camelia	*Camellia japonica*
	eurya	*Eurya japonica*
	Japanese skimmia	*Skimmia japonica*
	long stalk holly	*Ilex pedunculosa*
	false holly	*Osmanthus heterophyllus*
	coralberry	*Ardisia crenata*
	marlberry	*Ardisia japonica*
	gardenia	*Gardenia jasminoides*
	Japanese nutmeg tree	*Torreya nucifera*
	Japanese plum yew	*Cephalotaxus harringtonia*
	Japanese woodlander	*Ainsliaea apiculata*
Cool-Temperate (Montane Temperate) Broadleaf	albrechtii	*Rhododendron albrechtii*★
	Japanese beech	*Fagus crenata*★
	Japanese blue beech	*F. japonica*★
	white oak	*Quercus crispula*★
	tall stewartia	*Stewartia monadelpha*★
	snow camelia	*C. rusticana*★
	borealis aucuba	*A. japonica* var. *borealis*★
	clusterberry	*I. leucoclada*★
	Japanese ash	*Fraxinus lanuginosa*★
	summer snowflake	*Viburnum furcatum*★
	nettle-leaved hydrangea	*Hydrangea hirta*★
	maple	*Acer argutum, A. capillipes, A. carpinifplium, A. cissifolium, A. distylum, A. japonicum, A. micranthum, A. mono, A. nikoense, A. palmatum,* and *A. rufinerve*
Subalpine Coniferous		
Honshu and Shikoku only	Veitch's fir	*Abies veitchii*★
	maries fir	*A. mariesii*★
	hondo spruce	*Picea jezoensis* var. *hondoensis*★
	Japanese spruce	*P. bicolor, P. koyamae, P. maximowiczii, P. shirasawae*★
	northern Japanese hemlock	*Tsuga diversifolia*★

Table 7-2. Continued

Rainforest Zone	Common Name	Scientific Name
Honshu only	alpine mitrewort	*Menziesia pentandra*
	arrowhead rosettes	*Pteridophyllum racemosum*
	lousewort	*Pedicularis keiskei*
	nippon lily	*Coptis japonica* var. *dissecta*
	gold thread	*C. quinquefolia*
Hokkaido and Sakhalin only	Jezo spruce	*Picea jezoensis*
	Sakhalin fir	*Abies sachalinensis*
	Glehn's spruce	*P. glehnii*

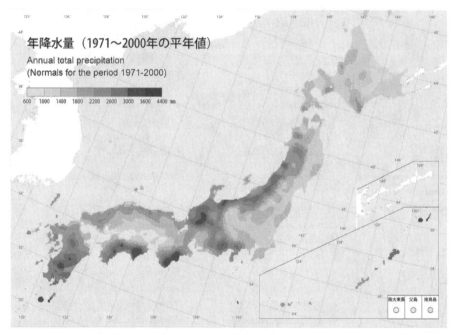

Figure 7-2. Annual precipitation from 1971 to 2000 for the Japanese archipelago, based on data from the Japan Meteorological Agency (2009).

latitude. During the monsoon, blowing winds deliver abundant precipitation, with some islands receiving heavy rainfall in the summer and others the winter.

Annual precipitation is high, with many places above 3,600 millimeters (see figure 7-2). With the exception of Hokkaido, the rainy season is from mid-June to July, and again in August to September during the typhoon season. Typhoons occur primarily on the Pacific side; however, during the winter season the Japan Sea side becomes snowbound due to the influence of westerlies coming from continental Siberia.

The archipelago's wet and mild oceanic climates saturate mountainous areas with enough moisture to sustain temperate rainforests (Miyawaki et al. 1980–1989; Krestov and Nakamura 2007). Meteorological data from Japan's temperate rainforest region can be summarized as follows: annual temperatures of 6–23°C, annual precipitation above 1,600 millimeters, and a continentality index (i.e., the range between the average temperature of the warmest [T_{max}] and coldest [T_{min}] months of the year) of 23–25°C.

Japan is split between warm (average temperature 13–23°C) and cool-temperate (average temperature 6–13°C) zones (see figure 7-1). The warm-temperate zone lies from Kyushu District (~30°N latitude) north to the southern Tohoku District (~39°N latitude), and is characterized by annual precipitation of 1,200 to 1,700 millimeters. Above 39°N latitude, warm gives way to cool-temperate, with some areas receiving more than 2,800 millimeters precipitation annually. However, the Japan Sea side receives much greater winter precipitation as snowfall than do Honshu and Hokkaido, which are on the Pacific side. In general, the cool-temperate zone is synonymous with per-humid conditions as described in Chapter 1.

RAINFOREST DISTRIBUTION

Because of the periodic influence of seasonal monsoons and high rainfall, most of Japan's temperate forests can be considered rainforest, although some rainforest species such as Siebold's beech (*F. crenata*) can also occur in the drier non-rainforest habitats of northern Japan (e.g., southern Hokkaido), where there is more winter precipitation (Matsui et al. 2004a). Japan's island ecology and diverse topography also contribute to its striking vegetation zonation from warm- to cool-temperate and subalpine and alpine that is latitudinally distributed. Thus, temperate rainforests are generally distinguished by their differences in temperature and moisture as described above for the two temperate zones, and by plant-species composition as follows.

Warm-Temperate (Lowland and Colline Belt) Rainforest

Oceanic climate, especially a warm current, extends the northern limit of evergreen forests into the warm-temperate zone. Here, warm-temperate rainforest, also known as the evergreen broadleaf zone or Laurisilva, occurs on Kyushu (~33°N latitude), Shikoku (~34°N latitude), and southern Tohoku of the Honshu mainland (~35°N latitude). Warm-temperate rainforests include many endemic plants (see table 7-2). But, in general, this zone has two types of

evergreen broadleaf forests: *Castanopsis-Persea*, occurring along the lower elevation belt, in particular along the sea coast area where the climate is mild; and
Quercus-Abies, occurring along the upper elevation belt.

Through the ages, intensive land use converted most of the *Castanopsis-
Persea* zone to rural landscapes, including rice paddies, agricultural fields, orange
gardens, coppice, and bamboo forests. Within this landscape is a symbolic landmark, Chinju-no-mori (the Shinto-shrine forest), which is protected as natural
or semi-natural rainforest. Wild animals such as the raccoon dog (*Nyctereutes
procyonoides*), the Japanese red fox (*Vulpes vulpes japonica*), the Japanese dwarf flying squirrel (*Pteromys momonga*), the Japanese giant flying squirrel (*Petaurista
leucogenys*), and the Ural owl (*Strix uralensis*) can be found in the remaining
Chinju-no-mori. However, many animals introduced as pets are now feral, including the North American raccoon (*Procyon lotor*), the masked musang
(*Paguma larvata*), and the Taiwan squirrel (*Callosciurus erythraeus thaiwanensis*),
which cause problems for native species.

The *Quercus-Abies* forest zone is characterized by evergreen oaks and a broad
mixture of flowering plants (see table 7-2). This region has been converted predominately to tea and fruit gardens, mulberry fields, coppice forests, and Japanese
cedar (*Cryptomeria japonica*)-Japanese cypress (*Chamaecyparis obtusa*) plantations.
The giant salamander resides here along middle and upper streams with good reserved catchment area. Notably, 18 of 19 salamander species are endemic, an indicator of clinal (environmental zonation along gradients) water conditions.

Cool-Temperate (Montane Temperate) Rainforest Belt

The cool-temperate rainforest zone is represented by the deciduous endemic
beech (*F. crenata*) forest above 1,200 meters elevation in Kyushu, at 1,000–1,800
meters in Shikoku, 800–1650 meters in Chubu, and in the lowlands of south
Hokkaido (Matsui et al. 2004a, plate 12a). North of Hokkaido, beech gives
way to white oak (*Quercus crispula*) and Sakhalin fir (*Abies sachalinensis*), but
these forests are considered too dry (1,000 millimeters annually) to qualify as
rainforest.

Beech rainforests include many endemics (see table 7-2; plate 12b). However, in Nagano and Yamanashi continental species such as Asian hazel (*Corylus heterophylla* var. *thunbergii*), Japanese tree lilac (*Syringa reticulata*), and Korean
pine (*Pinus koraiensis*) occur above the deciduous broadleaf forest belt at ~700
meters in place of beech. Above ~1,650 meters elevation, evergreen coniferous
rainforests are very similar to the subalpine forests of eastern North America
that occur on montane slopes (Franklin et al. 1979, Nakamura et al. 1994,
Nakamura and Krestov 2005). Podosolic soil (i.e., an acidic, heavily leached,

moist soil type) is characteristic of cold and wetter conditions that are well suited to coniferous species, including the genera *Abies, Picea, Tsuga, Larix,* and *Pinus.* In general, Japanese rainforest trees are shorter-lived and smaller in stature than those in similar rainforests on the Pacific Coast of North America (Franklin et al. 1979; Nakashizuka 1987). Along the Japan Sea side, coniferous forest cannot survive in the snowbelt. Prostrated birch forests and tall forbs occur on avalanche-prone slopes. The upper zone is therefore marked by Siberian dwarf pine (*P. pumila*), extending to 2,900 meters elevation; above this zone krummholz (tree line) appears. From here, the landscape is characterized by the spectacular Japanese Alps, carpeted by dwarf shrubs, meadows, and bare land.

Beech forest along the Pacific side is characterized by dwarf bamboos, such as *Sasa nipponica* and *Sasamorpha borealis,* and various mixtures of Japanese blue beech (*F. japonica*) and Siebold's beech. Several endemic species help further differentiate this rainforest type on the Japan Sea side (see table 7-2). Notably, similar beech forests occur in mainland Asia, Europe, and along the east coast of North America, although (except in Norway and Korea) they occur typically in drier and cooler non-rainforest climates (Ohba 1985; Miyawaki et al. 1994). Some isolated areas of beech occur in Taiwan and South Korea as well. These forests can have a diverse range of structures from tree mortality and delays in forest regeneration due to dense bamboo (*Sasa* spp.) understory and forest-management practices (Nakashizuka 1987; Masaki et al. 1999).

After World War II, plantations of Japanese cedar, false cypress, Japanese larch (*L. kaempferi*), and Japanese red pine (*P. densiflora*) expanded, while beech forest contracted quickly until today. However, the Shiragami Mountains in Tohoku are designated as a World Heritage Site, protected from logging. Here, beech rainforest is home to Asiatic black bear (*Ursus thibetanus*), badger (*Meles meles*), dormouse (*Glirulus japonicus*), black woodpecker (*Dryocopus martius*), golden eagle (*Aquilea chrysaetos*), and Japanese macaque.

Subalpine Rainforest

Ecologists generally do not recognize a boreal zone in Japan; however, coniferous forests that share some species with the Chinese or Siberian taiga, other environmental features with boreonemoral, and features with boreal or snow rainforests in other rainforest regions occur on upper montane and subalpine belts in several places. Examples of these forests can be found on central Honshu, the Chubu Mountains (1,650 to 2,650 meters elevation), northern Honshu, Tohoku (~40°N latitude, 1,200–2,200 meters elevation), and Hokkaido (~43°N latitude; 800–1,900 meters elevation). This upper belt is marked by

Siberian dwarf pine (*P. pumila*), extending to 3,000 meters elevation; above this zone krummholz (tree line) appears.

Different coniferous species are distributed on the Japanese islands within this zone (Nakamura and Krestov 2005; see table 7-2). In addition, Jezo spruce (*P. jezoensis*), found throughout the subalpine zone, has a wide distribution in East Asia, extending to Kamchatka, Russia. This is presumably because Hokkaido was connected with the continent through Sakhalin (Russian Far East) several times during ice ages.

In general, evergreen coniferous forest in this zone developed under rainy, oceanic influences. However, during the glacial period several circumpolar species invaded Japan's rainforests, including bunchberry dogwood (*Cornus canadensis*), wood sorrel (*Oxalis acetosella*), alph wood (*Maianthemum dilatatum*), sidebells wintergreen (*Orthilia secunda*), creeping lady's tresses (*Goodyera repens*), and heartleaf twayblade (*Listera cordata*).

THREATS

The primary threats to Japan's remaining temperate rainforests are ongoing losses from land conversion, climate change, and population increase and range expansion of Japanese deer and macaques. Both of these species are particularly hard on native plants as they can overgraze; climate change is allowing them to migrate upward in elevation. Overall, there is less anthropogenic influence in the upper montane and subalpine belts.

Western Japan's lowland areas were once occupied by native trees, including Japanese machilus (*Machilus thunbergii*), Japanese zelkova (*Zelkova serrata*), Japanese hackberry (*Celtis sinensis*), aphananthe (*Aphananthe aspera*), and Japanese alder (*Alnus japonica*). Until the Edo-period (1603–1868) there were many undisturbed natural forests in Japan's mountainous areas. The Meiji-Restoration government (1868–1912) protected forests as a kind of imperial national park. In Hokkaido, at the time, the Ainu (a group indigenous to Hokkaido, the Kuril Islands, and much of Sakhalin) used forests to cultivate wild food and make clothing materials while holding forests in great reverence. The increasing wealth and military power of Japanese people through the Meiji Era was accompanied by cultivation of most native trees, which continued through and after World War II.

In particular, after World War II a national policy of "expansive afforestation" destroyed many of Japan's natural beech forests. This occurred from 1950 to 1970, during which ~300,000 hectares of native forests annually were converted to plantations comprised of Japanese cedar, false cypress, and Japanese larch to aid in post–World War II reconstruction efforts (Forest Agency 2009). In central Japan, for example, while natural secondary forests dominated the landscape in 1947, fifty years later the landscape became extremely fragmented by conifer plantations, the dominant land use in the cool-temperate zone (Miyamota and Sano 2008). Much of this was enabled by Japan's Forest Agency, established after the Meiji to further the conversion of its native forests to plantations.

While Japan's forests now total some 25 million hectares of semi-natural forest, plantation, and second-growth forest, only about 7 million hectares (28 percent) is considered natural forest (Environment Ministry 1999). Moreover, the current system of parks and reserves does not adequately represent the range of environmental variation for critical rainforest elements (Kamei et al. 2006). Some 400,000 hectares (5.7 percent) of intact natural forest are protected as Natural Forest Ecosystem Conservation Areas in 27 of Japan's 47 prefectures[1] under the Protected Forest Institution by Japan's Forest Agency (Endo 2008; but see table 10-1 for minor differences in forest totals). Japan's Forest Agency also recently proposed long-term maintenance and restoration of broadleaf rainforest for carbon sequestration as part of the nation's commitment to climate-change mitigation (Forest Agency 2009). Some representative temperate rainforest (beech forest) remains in parks and religious shrines; however, it has contracted considerably and is not fully protected. The Shiragami-Sanchi World Heritage Site in northern Honshu also includes the last virgin remains of the cool-temperate rainforest of Siebold's beech trees that once covered the hills and mountain slopes of northern Japan. Black bear, Japanese serow (*Capricornis crispus* or *Naemorhedus crispus*, a small bovid) and 87 species of birds can be found in this forest.

The unique climate of Japan, located at the confluence of subtropical ocean currents and subarctic winds from Siberia, with most of its rain coming from Pacific monsoons, combined with both the evolutionary and the human-caused isolation of its species, make Japan quite susceptible to stresses created by global climate change. Models of climate change generally predict that here, as in many other temperate and sub-boreal oceanic climates, rainfall from monsoons will increase in intensity and duration, and winter precipitation and snow accumulation will decline. The unique subalpine vegetation will likely be lost

---

[1] The prefectures of Japan refer to the nation's 47 subnational jurisdictions. In Japan, they are commonly referred to as todōfuken and are made up of governmental bodies larger than cities, towns, and villages.

on all but the highest mountains in central Japan, and the cool-temperate beech forests are expected to disappear from all but the extreme northeastern parts of the country in Hokkaido (Matsui et al. 2004b). This will be principally caused by high temperatures during the growing season in the southern islands, and the lack of snow or decreases in winter precipitation from Honshu northward. The highly fragmented and humanized landscapes, and the insular nature of the Japan, will also make species' migrations to suitable habitats difficult at best.

Although Japan imports most of its wood-fiber needs, Japanese forests are reaching maturity and could be logged again if imports decline. Moreover, the Japanese government continues to expand its plantation system, to the detriment of native forests. In fact, this is the first time in Japanese history that plantation forests have outnumbered native forests. Like temperate rainforests around the world, the race is on to save what remains of these rainforests against the backdrop of global climate change and increasing natural-resource demands. Conservation efforts need to focus on slowing conversion of native forests by protecting and restoring remaining rainforests in their semi-natural state and providing landscape linkages for plants and wildlife to redistribute along elevation zones across the Japanese archipelago.

LITERATURE CITED

Box, E. O. 1995. Climatic relations of the forests of East and Southeast Asia. Pp. 23–55 in *Vegetation science in forestry.* Ed. E. O. Box, R. K. Peat, T. Masuzawa, I. Yamada, K. Fujiwara, and P. F. Maycock. Dordtrecht: Kluwer Academic Publishers.

Endo, K., ed. 2008. Modern forest policy. Japan Forestry Investigation Committee, Tokyo. (In Japanese.)

Environment Ministry 1999. The report of vegetation survey. The fifth natural environment conservation basic research. Nature Conservation Bureau of Environment Ministry and Asia Air Survey.

Forest Agency, ed. 2009. *Forest and forestry white book.* Japan Forestry Society, Tokyo. (In Japanese.)

Franklin, J. F., T. Maeda, Y. Ohsumi, N. Matsui, and H. Yagi. 1979. Subalpine coniferous forests of central Honshu, Japan. *Ecological Monographs* 49 (3):311–34.

Hämet-Ahti, L., T. Ahti, and T. Koponen. 1974. A scheme of vegetation zones for Japan and adjacent regions. *Annales Botanici Fennici* 11:59–88.

Japan Meteorological Agency 2009. Climatic atlas of Japan for the period of 1971–2000. Homepage of Japan Meteorological Agency. www.data.jma.go.jp/obd/stats/data/mdrr/atlas/precipitation/precipitation_13.pdf

Kamei, M., and N. Nakagoshi. 2006. Geographic assessment of present protected areas in Japan for representativeness of forest communities. *Biodiversity and Conservation* 15:4583–600.

Krestov, P.V., and Y. Nakamura. 2007. Climatic controls of forest vegetation distribution in Northeast Asia. *Berichte der Reinhold-Tüxen-Gesellschaft* 18:131–45.

Masaki, T., H. Tanaka, H. Tanouchi, T. Sakai, and T. Nakashizuka. 1999. Structure, dynamics, and disturbance regime of temperate broad-leaved forests in Japan. *Journal of Vegetation Science* 10:805–14.

Matsui, T., T. Yagihashi, T. Nakaya, N. Tanaka, and H. Taoda. 2004a. Climatic controls on distribution of *Fagus crenata* forests in Japan. *Journal of Vegetation Science* 15:57–66.

———, T. Yagihashi, T. Nakaya, H. Taoda, S. Yoshinaga, H. Daimaru, and N. Tanaka. 2004b. Probability distributions, vulnerability and sensitivity in *Fagus crenata* forests following predicted climate changes in Japan. *Journal of Vegetation Science* 15:605–14.

Miyamoto, A. and M. Sano. 2008. The influences of forest management on landscape structure in the cool-temperate forest region of Japan. *Landscape and Urban Planning* 86:248–56.

Miyawaki, A., ed. 1980. *Vegetation of Japan*. Band 1: Yakushima. Shibundo, Tokyo. (In Japanese with German summary).

———, ed. 1981. *Vegetation of Japan*. Band 2: Kyushu. Tokyo: Shibundo. (In Japanese with German summary.).

———, ed. 1982. *Vegetation of Japan*. Band 3: Shikoku. Tokyo: Shibundo. (In Japanese with German summary.)

———, ed. 1983. *Vegetation of Japan*. Band 4: Chugoku. Tokyo: Shibundo. (In Japanese with German summary.)

———, ed. 1984. *Vegetation of Japan*. Band 5: Kinki. Tokyo: Shibundo. (In Japanese with German summary.)

———, ed. 1985. *Vegetation of Japan*. Band 6: Chubu. Tokyo: Shibundo. (In Japanese with German summary.)

———, ed. 1986. *Vegetation of Japan*. Band 7: Kanto. Tokyo: Shibundo. (In Japanese with German summary.)

———, ed. 1987. *Vegetation of Japan*. Band 8: Tohoku. Tokyo: Shibundo. (In Japanese with German summary.)

———, ed. 1988. *Vegetation of Japan*. Band 9: Hokkaido. Tokyo: Shibundo. (In Japanese with German summary.)

———, ed. 1989. *Vegetation of Japan*. Band 10: Okinawa/Ogasawara. Tokyo: Shibundo. (In Japanese with German summary.)

———, K. Iwatsuki, and M. M. Grandtner, eds. 1994. *Vegetation in Eastern North America*. Tokyo: Univ. of Tokyo Press.

Nakamura, Y., M. M. Grandtner, and N. Villeneuve. 1994. Boreal and oroboreal coniferous forests of eastern North America and Japan. Pp. 121–54 in *Vegetation in Eastern North America*. Ed. A. Miyawaki, K. Iwatsuki, and M. M. Grandtner. Tokyo: Univ. of Tokyo Press.

———, and P.V. Krestov. 2005. Coniferous forests of the temperate zone of Asia. Pp. 165–220 in *Coniferous forests*. Vol. 6 of *Ecosystems of the world*. Ed. F. Andersson. New York: Elsevier Academic Press.

Nakashizuka, T. 1987. Regeneration dynamics of beech forests in Japan. *Vegetatio* 69:169–75.

Ohba, T. 1985. Beech forests of Japan and world. Pp. 201–30 in *The culture of the beech forest zone*. Ed. T. Umehara, T. Ichikawa, and T. Shidei. Tokyo: Shisakusha. (In Japanese.)

Qian, H., and R. E. Ricklefs. 2000. Large-scale processes and the Asian bias in species diversity of temperate plants. *Nature* 407:180–82.

Rohrig, E., and B. Ulrich, eds. 1991. *Temperate deciduous forests*. Vol. 7 of *Ecosystems of the world*. New York: Elsevier Academic Press.

Shimizu, T. 1983. *The new alpine flora of Japan in color II*. Tokyo: Hoikusha Publishing.

United Nations Development Programme, United Nations Environment Programme, World Bank, and World Resources Institute. 2000. World Resources 2000–2001: People and ecosystems: The fraying web of life. Washington, DC.

Yamanaka, T. 1979. *Forest vegetation of Japan*. Tokyo: Tsukijishokan. (In Japanese.)

CHAPTER 8

Temperate Rainforests
of Australasia

James B. Kirkpatrick and Dominick A. DellaSala

Temperate rainforests of western and northeastern Tasmania, mainland Australia (New South Wales, Victoria), and New Zealand, collectively referred to as Australasia throughout this book (see figure 8-1), are restricted primarily to coastal areas of latitudes 39.5 to 43.5°S latitude. These rainforests are valued for their beauty, mystery, and spirituality by nearly all Australians and New Zealanders, having been globally recognized by scientists, bureaucrats, and nongovernmental organizations for outstanding universal significance according to a wide variety of tests of the forests' importance.

RAINFOREST VITALS AND GLOBAL ACCOLADES

One test of the global importance of these remarkable rainforests is the preponderance of outstanding phenomena: big things or the most of a set of things. In particular, size matters for these rainforests, as they are home to several of the world's largest living or extinct taxa, including the largest living eagle (wedgetail eagle, *Harpagornis moorei*), the largest extinct eagle (Haast's eagle, *Aquila audax fleayi*), and the largest freshwater crustacean (tayetea, *Astacopsis gouldi*), which frequents rainforest and other streams in northwest Tasmania (see table 8-1). The extinct Haast's eagle once hunted extinct giant moas (*Dinomas robustus*) in New Zealand, while the Tasmanian subspecies of wedgetail eagle also frequents rainforests, though it finds most of its prey elsewhere. The largest

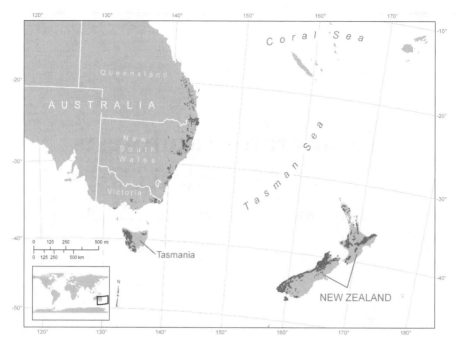

Figure 8-1. Temperate rainforests of Australasia, including the Australian coastline, Tasmania, and New Zealand (adapted from Kirkpatrick and Dickerson 1984).

Table 8-1. Globally significant attributes of temperate rainforests of Australasia.

Attribute	Importance
Ancient Gondwana affinities with many species unchanged for millions of years	Unique species assemblages and affinities shared with South African and Chilean temperate rainforests
Large intact areas	Unimpaired ecosystem and evolutionary processes, and high levels of ecological integrity
Trees up to 2,000 years old, rivalling California redwoods and Chilean *alerce* forests	Habitat for wildlife species dependent on old forests; globally significant carbon-dense forests
Unusually large organisms such as *Eucalyptus*, wedgetail eagle, and Tayeta	Uniqueness of place
Global "hotspot" for endemic taxa, including many vascular plants, invertebrates, amphibians, reptiles, birds, and mammals	Uniqueness of place
Mixed forests, treeless moorlands, and rainforest in close proximity	High levels of beta diversity (species turnover across environmental gradients)

carnivorous marsupials, both extinct (thylacine, *Thylacinus cynocephalus*) and living (Tasmanian devil, *Sarcophilus harrisii*), once frequented or now live in Tasmanian rainforests as well as other types of habitat. The largest flowering plant (mountain ash, *Eucalyptus regnans*) occurs in mixed forests in Victoria and Tasmania. If mixed forest is regarded as rainforest, as it should be, mountain ash, which occurs in the temperate rainforests of Tasmania and Victoria, is a giant among other giants, towering to over 100 meters. Finally, the largest number of shrub species with a filiramulate (divaricate) structure (i.e., slender interlacing twigs with small distantly clustered leaves) is found in the temperate rainforests of New Zealand (Wardle 1991).

A second test of global importance is uniqueness, as applied in defining the global biodiversity "hotspots" of Conservation International (see Myers et al. 2000). Are there a lot of species found nowhere else in the world? The answer is a resounding yes for New Zealand rainforests, which are characterized by numerous endemics. The same pertains to the Tasmanian and Australian temperate rainforests considered as one, although many more species than in New Zealand occur widely in other vegetation types. There is also a very high degree of Tasmanian endemism within its thamnic and implicate rainforest biota (see below for rainforest types). In all cases, endemism is highest among the invertebrates and vascular plants and least among the bryophytes. Many species of mammals, birds, amphibians, and reptiles that can be seen in these rainforests are found nowhere else in the world outside Australia, including the Tasmanian devil, the eastern quoll (*Dasyurus viverrinus*), and the platypus (*Ornithorhynchus anatinus*). Ancient species such as the mountain shrimp (*Anaspides tasmaniae*) and beech (*Nothofagus* spp.) have changed little over millions of years.

Temperate rainforests of Australia and New Zealand also provide universally outstanding examples of major ecological and evolutionary processes at work. Complex dynamic relationships among soil, fire, climate, and biota in southwestern Tasmania are expressed in a landscape with widespread treeless moorlands made up of fire-dependent buttongrass (*Gymnoschoenus sphaerocephalus*) mixed with a variety of rainforest types, including mixed forest. Such richness has underscored efforts to designate portions of the region among World Heritage sites. In fact, the large number of "primitive" temperate rainforest species, including many with strong Gondwana affinities, was accepted as a major argument for listing all four of the World Heritage Areas in Australia and New Zealand that have this type of vegetation. Indeed, much of the surviving temperate rainforest in Australasia occurs in areas with wilderness qualities and high relative relief, which increases its value for World Heritage status. Last but not least, individual rainforest pine-relatives can live to 2,000 years, with some clonal tree complexes estimated at over 10,000 years.

In sum, the temperate rainforests of Australasia are a global conservation priority as recognized by Conservation International (global "hot spot"—Myers et al. 2000), the World Wildlife Fund (Global 200 ecoregion, Olson and Dinerstein 1998), and World Heritage status. Like the British Columbia and Alaska rainforests, some of the largest contiguous temperate rainforest blocks in the world occur here (e.g., Tarkine Wilderness in northwestern Tasmania and the western portion of the South Island of New Zealand).

ANCIENT AFFINITIES

About 180 million years ago, a single giant supercontinent, Pangea, is believed to have broken apart into Laurasia (now North America and Eurasia), and Gondwana (now South America, Africa, India, Antarctica, and Australia). Consequently, many rainforest species share close ties with species in the temperate rainforests as far away as Chile and South Africa (see chapters 5, 9). Although overlaying the South Pole, Gondwana had substantial forest cover. It eventually fractured, with fragments drifting in different directions to the north, leaving the once lushly forested Antarctica to become an almost lifeless expanse of ice caps, glaciers, and ice shelves.

About 45 million years ago, the Australian plate began its journey from Antarctica towards collision with the Asian plate, an impact that created the high mountains of New Guinea. Gondwana trees, such as the southern beeches (*Nothofagus*), were "life-boated" by the Australian plate to these highlands. Other plates also shipped Gondwana species to the north. The descendants of the inhabitants of the Gondwana forests best survive in the southernmost parts of the Gondwana fragments, such as Patagonia, southern Chile, New Zealand, and Tasmania (also see chapter 5).

For much of its journey north, Australia was Amazonian in the extent of its rainforest, which covered even the heart of the continent (White 1986). Giant rivers, their anabranches and lagoons, and the red, yellow, and white horizons of laterite soils can still be seen in the desert landscape, while a few survivors of the rainforest, like fan palms (*Livistona* spp.) and figs (*Ficus* spp.), sit among red rocks in deep gorges that receive constant seepage.

As Australia drifted into the zone of subtropical meteorological highs, the other fragments of Gondwana configured themselves to allow a circumpolar current, and land masses became concentrated around the North Pole. The planet started to flicker between brief periods of warmth and wetness and longer periods of coolness and aridity. We now live in one of the warm and wet periods.

Only 18,000 years ago, at the end of the last ice age, in the cool-temperate parts of Australia and New Zealand temperatures were approximately 6°C cooler and precipitation about half that at present. This was the last of many glacial periods in which rainforests were forced into refugia or local extinction. In Tasmania and Victoria, fossils deposited before the onset of these glacial periods attest to rainforest much richer in tree species than present (Hill et al. 1999). In New Zealand, diverse forests physiognomically similar to those once found in Tasmania can be seen in the southeast and central west of the South Island, and well south of the southernmost point of mainland Tasmania. Most of the rainforests of the South Island, however, are species-poor in vascular plants compared to those of Tasmania. South Island temperate rainforests are almost entirely comprised of *Nothofagus* that have proven adept at rapidly recolonizing ground bared of forest by Pleistocene glaciers and aridity.

Given this history, it is not surprising that the rainforests of the region are rich in species with Gondwana affinities. Many of these are regarded as primitive in that they have changed little morphologically, many since the days of the dinosaurs, and appear to be the ancestors of more-modern species. It is also not surprising that, at the level of genus, there is much in common among the rainforests of Tasmania, New South Wales, New Zealand, high mountain areas of New Guinea (tropical, but the high-latitude elements are Gondwana), New Caledonia (tropical), and Valdivia. The species that dominate the remnants of the Gondwana forest have evolved in response to the different environmental histories of these different land masses. However, many species in the same genus, like myrtle beech (*Nothofagus cunninghamii*) in Australia, silver beech (*N. menziesii*) in New Zealand and Magellan's beech (*N. betuloides*) in Chile, are as similar as nonidentical triplets.

The differences between the rainforests of the different land masses seem to have been as much influenced by extinction as adaptive radiation (that is, rapid evolutionary events). The extant Chilean rainforest genus, *Fitzroya*, has been only found as a fossil in Tasmania. Remaining relatively unchanged since the Triassic (some 230 million years ago), the mountain shrimp survived in Tasmania, but nowhere else. The tree genus *Eucryphia* is found on mainland Australia, Tasmania, and South America, but not in New Zealand. *Fuschia* is found in New Zealand and South America, but not Australia or Tasmania. Tasmania and South America have marsupial faunas. New Zealand has no native marsupials. Some geologists have suggested that New Zealand submerged at one stage, to be recolonized long distance from Australia after emergence. Its lack of any native nonmarine, non-flying mammals is consistent with this hypothesis.

EMERALD ISLES IN A SEA OF FIRE

Throughout most of Tasmania, Victoria, New South Wales, and Queensland, rainforest was scattered as islands of emerald in a sun-scorched sea of eucalypt forest and woodland. In contrast, when the Maori arrived in New Zealand, less than 1,000 years ago, the rainforest was the "sea," and the scorched islands were too cold for trees, or were recently defoliated by fires associated with volcanic eruptions. Fires set by Maori themselves eliminated rainforest from large areas in the drier parts of New Zealand (Wardle 1991). The arrival of the Maori also resulted in the extinction of the eagle and moa. Land clearing that followed the European invasion of the nineteenth century further depleted rainforest, preparing the way for weeds and pests.

At time of the European migration, a little more than 200 years ago, the largest contiguous patch of Australasian rainforest (~200,000 hectares) was in the Tarkine region of northwest Tasmania. This forest has largely survived to the present, unlike the less extensive rainforests that then occupied the fertile alluvial valleys of northern New South Wales, which were almost totally cleared for sugar production and pastures for dairy cows. The Australian rainforests had been inhabited and utilized by the Aboriginal people, who migrated to Australia more than 30,000 years ago. Their management of the landscape, using fire as a tool, may have influenced the distribution of rainforest and seems to have been responsible for the extinction of much of the striking Pleistocene marsupial megafauna depicted in cave paintings.

Despite these impacts, temperate rainforests of Australia and New Zealand still support an amazingly diverse range of vegetation types, characterized by large numbers of species of animals and endemic plants. In Tasmania, nearly three-quarters of the old-growth rainforest present before European arrival has been spared from logging, fire, industrial pollution, and other developments. Approximately one-third of New Zealand is still covered by temperate rainforest and native vegetation returning to rainforest, despite substantial attrition from clearing, fire, and logging. The outstanding universal significance of these rainforests partly relates to their beauty, but mostly to the many unusual and ancient species that illustrate major stages in the geological and evolutionary history of the Earth.

RAINFOREST CLIMATE

New Zealand and Tasmania are entirely in the temperate zone, superficially leaving no doubt that all their rainforests are "temperate." However, in northern Queensland, Australia, it is possible in a long day to walk up Mount Bellenden Kerr, commencing in a complex evergreen tropical rainforest, with profusions of palms and vascular epiphytes, and lianas worthy of use by Tarzan, and finishing in a decidedly cool-temperate rainforest with an endemic leatherwood (*Eucryphia* spp.) glowing green with mosses. Nature has put out no signs to mark the point at which tropical becomes temperate, or warm becomes cool-temperate. The variation in species composition and physiognomy is more or less continuous, except where there are sudden breaks in geological conditions, which, of course, are largely independent of variation in climatic conditions.

In general, temperate rainforests of the region are distinguished by average annual rainfall of 1,300–3,500 millimeters (some areas receiving up to 6,300 millimeters, temperatures of 5–12°C, effective summer monthly rainfall of at least 40 millimeters, and areas where snow, fog, and frost are common (see Williams 1974; Harris et al. 1995; worldwildlife.org[1]). The wide range of temperatures means climatic conditions, in general, correspond to globally based definitions ranging from perhumid (or cool, as used here) to warm-temperate, as described in Chapter 1.

RAINFOREST CLASSIFICATIONS

Some authors of rainforest classifications in Australia make distinctions between cool-temperate, warm-temperate, subtropical, and tropical rainforest at a coarse level (Adam 1992), but subtropical rainforest includes extensive areas of forest in the warm-temperate zone, forests that closely resemble much of the more complex New Zealand rainforest. This sort of classification problem does not arise in most parts of the globe, where deserts, savannahs, or grasslands conveniently separate the "tropical" and "temperate" rainforests (e.g., see Valdivian rainforest in chapter 5). However, this continuity is part of the reason for the World Heritage status of the rainforest of northern New South Wales and southern Queensland at the far eastern extremity of the continent, where the subtropical to cool-temperate transition can be found in many areas, such as the

---

[1]www.worldwildlife.org/wildworld/profiles/terrestrial/aa/aa0413_full.html

Lamington (Queensland), Border Ranges (Queensland) and New England (South Wales) national parks.

With increasing latitude and altitude in both Australia and New Zealand, there is a general tendency for the understory to change in dominance from vine to fern to bryophyte (Webb 1978). However, the major distinction between temperate rainforest types in both New Zealand and Australia are forests with *Nothofagus* +/− *Atherosperma* +/− *Athrotaxis* versus those without. In Australia, such forests are almost entirely coincident with the concept of "cool-temperate," and obligingly occur at higher latitudes and altitudes than do the other temperate rainforests. In New Zealand, the pattern of distribution of forests with *Nothofagus* (*Atherosperma* and *Athrotaxis* are absent) and those without defies anything but a historical explanation, with a huge gap in *Nothofagus* on the west coast of the South Island comprehensively occupied by other rainforest tree species (Wardle 1991).

The forests with *Nothofagus* tend to be very species-poor in trees and other vascular plants, although rich in bryophytes, fungi, and lichens. In Australia, the cool-temperate rainforests richest in vascular plants are those on the poorest-quality sites for tree growth, in contradiction to the tendency in warm-temperate, subtropical, and tropical rainforests. This may be because glacial refugia for rainforest in Tasmania were so restricted (Kirkpatrick and Fowler 1998) that only the tree species capable of occupying the poorest sites survived. Most of these sites would have been alpine. Not surprisingly, almost all Tasmanian rainforest trees have been recorded as shrubs in alpine vegetation (Kirkpatrick 1997).

Cool-Temperate

In general, there are four main structural/floristic types of cool temperate rainforests recognized for Australia, New Zealand, and Tasmania: callidendrous, thamnic, implicate, and montane (Brown and Read 1996; see figure 8-2). Montane forest, recognized by others (Brown and Read 1996), is usually an open woodland type and was not considered here as temperate rainforest.

CALLIDENDROUS

The cool-temperate rainforests on the sites with the greatest growth potential are cathedral-like with open understories (Brown and Read 1996). They are usually dominated by *Nothofagus* species or sassafras (*A. moschatum*), the latter only in Australia and Tasmania. The Australian cool-temperate rainforests are almost all cathedral-like, usually dominated by myrtle beech (*N. cunninghamii*) or

Callidendrous Rainforest

Thamnic (and Gallery) Rainforest

Implicate Rainforest

Montane Rainforest

Figure 8-2. Rainforest profiles of four main rainforest types in Australasia. Source: Brown and Read (1996); reprinted with kind permission of Springer Science and Business Media.

sassafras in Victoria and by southern beech (*N. moorei*) in New South Wales and southern Queensland. In New Zealand, these cathedral-like forests are the norm, usually dominated by hard beech (*N. truncata*), red beech (*N. fusca*), silver beech or black/mountain beech (*N. solandri*).

However, there are also extensive forests where podocarps (evergreens in the Southern Hemisphere of the genus *Podocarpus* that have a pulpy fruit with one hard seed) and cypresses emerge from a closed canopy of southern beeches. The major such gymnosperms (vascular plants lacking seeds produced in ovaries) that share these forests with the beeches are rimu (*Dacrydium cupressinum*), miro (*Prumnopitys ferruginea*), kahikatea (*Dacrycarpus dacryoides*), Hall's totara (*Podocarpus hallii*), totara (*P. totara*), kaikawaka (*Librocedrus bidwillii*), and silver pine (*Lagarostrobus colensoi*) (Wardle 1991).

THAMNIC

In Tasmania, there are large areas of cool-temperate rainforest labeled as thamnic (Jarman et al. 1999). These forests have many shrubs and small trees in the understory and a much more varied and diverse tree stratum than do the callidendrous forests.

IMPLICATE

On sites with the poorest growth potential, callidendrous forests give way to implicate forests, a tangled mass of non-vertical tree stems through which it is nearly impossible to force passage. One tree species, aptly named horizontal (*Anodopetalum biglandulosum*), has a propensity for its thin stems to fall horizontally and then to resprout vertically, doing this repeatedly. Some have claimed to have unknowingly camped on horizontal thatch tens of meters above the hidden ground.

Both thamnic and implicate rainforests of Tasmania are noted for their Tasmanian endemic podocarps and cypresses. Several of these gymnosperms have been shown to live to be older than 1,000 years, with the huon pine (*Lagorastrobus franklinii*) having single stems older than 2,000 years, and clonal complexes possibly older than 10,000 years, such as the high-altitude male clonal group near Lake Johnson in western Tasmania. The two *Athrotaxis* species, King Billy pine (*A. selaginoides*) and pencil pine (*A. cupressoides*), are known to be almost as long-lived, with the latter also reproducing through root sprouts. The celery top pine (*Phyllocladus aspleniifolius*) and Cheshunt pine (*Diselma archeri*) complete the complement of rainforest tree gymnosperms. Among the angiosperm (flower-producing) trees, myrtle beech is common in both thamnic and implicate forests. A congenor, the deciduous beech (*N. gunnii*), dominates some small areas of implicate rainforest. This winter deciduous species, the sole southern beech of Australia and New Zealand, is often mixed with a bizarre, palmlike tree heath, pandani (*Richea pandanifolia*). Angiosperm trees common in thamnic and/or implicate rainforests are leatherwood (*Eucryphia lucida*), dwarf leatherwood (*E. milliganii*), sassafras, white waratah (*Agastachys odorata*), waratah (*Telopea truncata*), and native plum (*Cennarhenes nitida*), the latter three belonging to the Proteaceae, a family that links South Africa and Australia.

Warm-Temperate Rainforests

On the mainland of Australia, the types of forest that have been labeled "warm-temperate" are associated more with poor soils than with a particular climatic zone (Adam 1992). They have trees that lack buttresses, occur in pole-like assemblages, and contain fewer species than the adjacent subtropical rainforest on more-productive soils. Some prominent canopy dominants are coachwood (*Ceratopetalum apetalatum*), Australian white birch (*Schizomeria ovata*), yet another "sassafras," *Doryphora sassafras*, and lilly-pilly (*Acmena smithii*).

Poor soils are less common in tectonically active New Zealand than in long stolid Australia. Thus, most of the forests without *Nothofagus* would be classified as "subtropical" in the Australian classification, even though they extend well to the south of Tasmania. These New Zealand "warm-temperate" rainforests have been labeled conifer/broadleaf forests by Wardle (1991), the conifers being largely podocarps and cypresses. The typical structure is long-lived conifers emergent above a closed layer dominated by many species of shorter-lived broadleaf angiosperms (Ogden and Stewart 1995). On the northern part of North Island, New Zealand, these forests are dominated by giant kauris (*Agathis australis* in the Araucariaceae) and have many lianas, vascular epiphytes, tree ferns, and even the occasional palm. In the far south on the South Island, New Zealand, markers of the Australian "subtropical" rainforest such as vascular epiphytes are prominent (Wardle 1991).

FOREST SUCCESSION AND RAINFOREST DYNAMICS

The temperate zone of Australia consists of forests dominated mostly by the pastel-leaved and sparse-crowned eucalypts (*Eucalyptus* sensu lato), largely dependent on fire for regeneration and encouraging fire's incidence through its flammability. In places where fire finds it difficult to penetrate because of moist climates, rocky topography, or low rates of fuel accumulation associated with fertile soils, eucalypt forest gives way to closed-canopy rainforest (Bowman 2000). But, in general, the stereotypical rainforest succession story is relay floristics (i.e., integrated communities replacing one another much like runners in a relay race) after displacement of the original rainforest by fire, windthrow, mass movement, flood, volcanic ash, lava flows, or human disturbance.

The relay floristic concept involves the initial invasion of plants after large or intense disturbance by fast-growing but short-lived species well-suited to open, competition-free ground. The growth of these species creates environmental conditions that allow a set of longer-lived, more shade-tolerant rainforest species that do not require disturbance for successful regeneration in order to establish. These primary species eventually replace the colonizers that cannot regenerate under the shade of rainforest canopies. The colonizing trees are commonly regarded as eucalypts (*Eucalyptus* sensu lato) in Australia and

Tasmania, and manuka (*Leptospermum scoparium*) and kanuka (*Kunzea ericoides*) in New Zealand. Rainforest has been assumed to be ultimately stable vegetation, in that the death of any one individual tree is compensated by a new individual arising in the gap created by this or another tree's death. There is no place for eucalypts, kanuka, or manuka in a real rainforest. Or is there?

This is more than an academic question in Australia and Tasmania, where mixed forests, consisting of eucalypts up to 95 meters tall, tower over closed-canopy rainforest often over 40 meters tall (see plate 13ab). The mixed forests have been the focus of ongoing bitter conservation-versus -development debates (see below), whereas there is an almost unanimous acceptance of the desirability of complete protection of rainforest. There are two critical questions: To what extent are the successional stages of a vegetation type part of that vegetation type? Where should the boundaries be placed between stages? In countries like Australia and New Zealand where rainforest, but not necessarily other forest, is sacrosanct, these questions have considerable conservation import.

In reality, not all vegetation unanimously regarded as rainforest is old growth, or has passed through the successional stages described above. For instance, gaps in stands of large pencil pines may be colonized by King Billy pines and hybrids between pencil and King Billy pines rather than either species alone (Cullen and Kirkpatrick 1988). This relay successional process can take over 1,000 years before King Billy pine takes over.

Also, when fire penetrates rainforest, not all rainforest trees necessarily die. In Tasmania, some trees, like horizontal and high-altitude ecotypes of myrtle beach, resprout vigorously from basal burls. Others, like sassafras (*A. moschatum*), produce epicormic shoots consisting of sprouts that emerge from beneath the bark on their trunks. The unfortunate few, constituting all the gymnosperms and deciduous beech, are killed by fire and can only return via the dispersal of seed from unburned stands. With the exceptions of celery top pine, which is dispersed by birds, and huon pine, which has a seed that floats, their dispersal ability is highly limited (Gibson et al. 1995). Thus, much of the Tasmanian rainforest consists of even-aged resprouts and seedlings of the more fire-resistant rainforest species. Eucalypts are only in the mix if they were close by when the fire went through, as their normal dispersal distance is but twice tree height. In New Zealand, widespread death of rainforest trees as the result of disturbance is also often followed by even-aged rainforest tree regeneration (Stewart and Veblen 1982).

The temperate rainforests of Australasia are subject to a variety of types of dieback, some of which appear to be part of the pre-European systems, others that appear to be caused by organisms introduced by Europeans, and still others that have a more complex etiology. The major known causes of rainforest tree dieback are canopy exposure following breakage by wind or snow, damage to root systems, drought, introduced mammalian herbivores, outbreaks of native herbivorous insects, changes to soil hydrology, and introduced and native fungal pathogens. Tree dieback appears much more common in the rainforests of New Zealand than Tasmania or Australia, possibly because of European introduction of mammals, including an omnivorous marsupial, the brush-tail opossum (*Trichosorus vulpecula*), which naturally frequents the Tasmanian and Australian forests.

Dieback in the rata-kamahi (*Weinmannia racemosa–Metrosideros umbellata*) forests of the unstable, fertile western slopes of the Southern Alps of New Zealand has been proximally caused by opossum browsing. The badly affected stands tend to have trees older than 300 years, which make great nesting hollows for opossums (Stewart and Veblen 1982). Opossums do not cause dieback in the same forest type of the same age on nearby less-fertile granitic soils, where other species palatable to opossums are less abundant. If deer browse in the dying forests, regeneration of dominant trees is prevented (Wardle 1991).

Lest the reader gain the impression that these rainforests are in a state of collapse induced by their ecological fragility, it is worth giving the counteracting example of dieback in eucalypts induced by the invasion of rainforest into the understory. This widespread phenomenon may be due to the effects of a closed rainforest understory on the temperature regimes and/or fungal communities of the soil. Whatever the cause, dieback in the eucalypts can be reversed by felling the rainforest understory (Ellis et al. 1980). However, this action is now likely to bring social disapprobation.

THREATS

Primary threats to these rainforests include logging of mixed forests and invasive species, particularly prolific in areas intensively utilized by people. Climate change also now threatens to change rainforest dynamics by creating more frequent drought cycles, leading to more fire and replacement of rainforest communities by fire-adapted earlier successional stages.

With the major exception of temperate rainforests unfortunate enough to have emergent eucalypts, logging and clearing of the vegetation type had effectively ceased in Australia and New Zealand by the turn of the century, the public finding the demise of rainforest as unacceptable as that of whales. Despite massively funded campaigns by developers to maintain their possession, conservation groups and the public convinced politicians to stop old-growth rainforest logging and broad-scale clearance after a grueling quarter-century campaign. In 2007, nearly all temperate rainforest, except mixed forest in Australia, and all temperate rainforest on public land in New Zealand, were dedicated to conservation. There is no native-forest logging on public land in New Zealand, and very little native forest on private land.

The powerful native-forest logging lobby in Australia has concentrated its activity and substantial financial resources on maintaining access to highly profitable eucalypt wood chips from the most productive of eucalypt forests, these largely being the ones that are the penultimate stage of rainforest succession, the mixed forests. The successive expansions of forest reserves in Australia and Tasmania, as one campaign or election followed another, eventually swept up virtually all temperate rainforest without eucalypts, but only a moderate proportion of the mixed forests. Most of the mixed forest area is in Tasmania, the most pro-logging of the Australian states, where it is difficult to perceive the boundaries between forest industry, state government, and forest bureaucracy, and impossible to separate their aims. In Tasmania, at the turn of the millennium, 406,144 hectares out of an existing 597,406 hectares (68 percent) and an estimated pre-European area of 803,400 hectares (51 percent) of rainforest lacking eucalypts was in reserves, increased from 11 percent and 8 percent, respectively, in 1970 (Mendel and Kirkpatrick 2002). A large proportion of the remaining unreserved rainforest, including all of the Tarkine, was reserved in a later expansion in 2004, for which exact figures are unavailable. The pre-European wet eucalypt forest of Tasmania, which is almost all potentially rainforest of some type, covered 1,289,300 hectares. In 1970, 3.7 percent of the pre-European rainforest and 5.5 percent of the extant rainforest was reserved. In 2002, 22 percent of the original rainforest and about 32 percent of the surviving rainforest was preserved (Mendel and Kirkpatrick 2002). This reserve expansion was not accepted as sufficient by a large proportion of Tasmanians and most Australians. Most conservationists can see the virtues of wood production, but want wood to be either an agricultural crop or be confined to plantations and native forests that already have been cut. They see the existing plantation estate as more than sufficient to provide the wood that the nation needs, but those who make enormous profits from wood-chip exports seem to think otherwise (Ajani 2007).

Assuming that temperate rainforest without eucalypt emergents remains socially sacred, and therefore protected, the major management problems after fire exclusion relate to species that have been introduced as a consequence of the human migration of Australia and New Zealand and continue to be introduced as the direct result of unsustainable levels of international and national travel and trade. Fox (*Vulpes vulpes*), wild pig (*Sus scrofa*), and many introduced plants degrade rainforest or threaten native rainforest biota on mainland Australia. The invasion of the chytrid fungus (*Batrachochytrium dendrobatidis*) threatens frog populations everywhere. The root-rotting cinnamon fungus (*Phytophthora cinnamomi*) threatens rainforest species on poor soils below 800 meters elevation in Tasmania (Podger and Brown 1989), where there are no wild pigs, but where the fox is struggling to establish at the same time as Tasmanian devil populations are plummeting as the result of a new bite-transmitted cancer. In New Zealand, many native-forest songbirds may slide toward extinction, their eggs eaten by arboreal opossums along with the foliage of many trees, while deer eat seedlings on the forest floor.

While it seems unlikely that any of the existing, or prospective, invasions will lead to the demise of temperate rainforest as a biological community, they may cause widespread local extinctions, and, perhaps, total extinctions, of some temperate rainforest species. The conservation challenge is to effectively use very limited resources to control invasive species in the places in which such control is likely to yield maximum benefit for threatened species. Simple but difficult actions, like closure of human access to large forest wilderness areas that have yet to be invaded by introduced diseases and pathogens, need to be taken. The shooting of introduced animals and the use of 1080 poison, a potent metabolic poison used as an anti-herbivore metabolite, are options for controlling invasive species where these measures do not threaten native species. However, genetic engineering to produce diseases that sterilize invasive pests is risky, as a pest in either Australia or New Zealand is a precious native elsewhere, and it is difficult to see how any disease could be effectively confined to one country.

We should also not be too chauvinistic in our biodiversity conservation. The parma wallaby (*Macropus parma*), a species that used to be common on the margins of temperate rainforest in New South Wales, but is now rare and endangered by the fox in its natural habitat, was introduced to some smaller islands in amazingly fox-free New Zealand, where it eats some rainforest seedlings. Preventing the local loss of a few rainforest trees is surely not worth the further endangerment of a mammal of great beauty and interest.

WHAT WILL IT TAKE TO SAVE AUSTRALASIA'S RAINFORESTS?

There are excellent practical reasons for conserving the temperate rainforests of Australasia, not the least of which is their major role as a carbon warehouse. Much of the rainforest of western Tasmania grows on 1–2 meters of red fibrous peat sitting on quartzite. Large amounts of carbon are trapped in individual kauri and giant eucalypts, their understory, litter, and soils, and the carbon released by logging will never return if the forests are logged again. Unlogged forests stabilize slopes and trap and slowly release scarce water, used downstream in cities for irrigation and hydroelectric production. In most areas their value for maintenance of water supplies far exceeds that of the wood products they can generate (Creedy and Wurzbacher 2001). In fact, logging usually returns less money than the long-term value of water lost by replacing old forests with regenerating ones, which release much more water in transpiration than old forests.

Mitigation of climate change requires a cessation of logging in old-growth forests, which is presently a significant source of carbon dioxide emissions globally (IPCC 2007). Old-growth mixed forests of Australia and Tasmania, in particular, are among the world's most carbon-dense forests (on average 1,867 tonnes of carbon per hectare—Keith et al. 2009). It is probably not a good idea to liberate this carbon to produce glossy magazines and floorboards when the same materials can be produced from Monterey pine (*Pinus radiata*) and eucalypts from existing plantations.

Climate change may already be affecting the temperate rainforests of Australia and New Zealand by increasing the incidence of severe drought and fire. While little can be done about severe drought other than to watch species slowly retreat from north-facing to south-facing slopes, fire management is a necessity in Tasmania and Australia, where a lack of burning of the highly flammable vegetation types dominated by eucalypts or sedges juxtaposed to rainforest makes these forests more likely to burn. A reinstatement of indigenous-style burning of the lowland treeless vegetation has been suggested as a way to reduce the likelihood of further loss of rainforest to wildfire (Marsden-Smedley and Kirkpatrick 2000).

The world is fortunate that a large part of these rainforests survives in landscapes of striking beauty. We are doubly fortunate that people have worked hard to successfully ensure that most remaining old-growth rainforest is almost entirely protected from logging and land clearance, but we need to work harder so that we can soon say the same about the remaining spectacular mixed forests. We also have an ongoing challenge in ensuring the survival of those species threatened by human-induced climate change and an unprecedented mixing of biotas of the Earth.

LITERATURE CITED

Adam, P. 1992. *Australian rainforest*. Oxford: Clarendon Press.
Ajani, J. 2007. *The forest wars*. Carlton: Melbourne Univ. Press.

Bowman, D. M. J. S. 2000. *Islands of green in a sea of fire.* Cambridge: Cambridge Univ. Press.

Brown, M. J., and J. Read. 1996. A comparison of the ecology and conservation management of cool-temperate rainforest in Tasmania and the Americas. Pp. 320–41 in *High-latitude rainforests and associated ecosystems of the west coast of the Americas.* Ed. R. G. Lawford, P. B. Alaback, and E. Fuentes. Berlin: Springer-Verlag.

Creedy, J., and A. D. Wurzbacher. 2001. The economic value of a forested catchment with water, timber, and carbon sequestration benefits. *Ecological Economics* 38:71–83.

Cullen, P. C., and J. B. Kirkpatrick. 1988. The ecology of *Athrotaxis* D. Don (Taxodiaceae) II. The distributions and ecological differentiation of *A. cupressoides* and *A. selaginoides. Australian Journal of Botany* 36:561–73.

Ellis, R. C., A. B. Mount, and J. P. Mattay. 1980. Recovery of *Eucalyptus delegatensis* from high-altitude dieback after felling and burning the understorey. *Australian Forestry* 43:29–35.

Gibson, N., P. C. J. Barker, P. J. Cullen, and A. Shapcott. 1995. Conifers of southern Australia. Pp. 223–51 in *Ecology of the southern conifers.* Ed. N. J. Enright and R. S. Hill. Carlton: Melbourne Univ. Press.

Harris, S., J. Balmer, and J. Whinam. 1995. Western Tasmanian Wilderness. Pp. 495–99 in *Asia, Australasia, and the Pacific.* Vol. 2 of *Centres of Plant Diversity.* Ed. S. D. Davis, V.H. Heywood, and A.C. Hamilton. Cambridge, UK: WWF/IUCN, IUCN Publications Unit.

Hill, R. S., R. S. MacPhail, and G. J. Jordan. 1999. Tertiary history and origins of the flora and vegetation. Pp. 39–63 in *Vegetation of Tasmania.* Supplementary Series No. 8 of *Flora of Australia* Ed. J. B. Reid, R. S. Hill, M. J. Brown, and M. J. Hovenden. Canberra: ABRS.

Intergovernmental Panel on Climate Change (IPCC). 2007. Synthesis report. Geneva, Switzerland. www.ipcc.ch/contact/contact.htm

Jarman, S. J., G. Kantvilas, and M. J. Brown. 1999. Floristic composition of cool-temperate rainforest. Pp. 145–59 in *Vegetation of Tasmania.* Supplementary Series No. 8 of *Flora of Australia.* Ed. J. B. Reid, R. S. Hill, M. J. Brown, and M. J. Hovenden. Canberra: ABRS.

Keith, H., B. G. Mackey, and D. B. Lindenmayer. 2009. Re-evaluation of forest biomass carbon stocks and lessons from the world's most carbon-dense forests. *Proceedings of the National Academy of Sciences* 106 (28):11635–40.

Kirkpatrick, J. B. 1997. *Alpine Tasmania: An illustrated guide to the flora and vegetation.* Melbourne: Oxford Univ. Press.

———, and M. Fowler. 1998. Locating likely glacial refugia in Tasmania using palynological and ecological information to test alternative climatic models. *Biological Conservation* 85:171–82.

Marsden-Smedley, J., and J. B. Kirkpatrick. 2000. Fire management in Tasmania's Wilderness World Heritage Area: Ecosystem restoration using Indigenous-style fire regimes? *Ecological management and restoration* 1:195–203.

Mendel, L. C., and J. B. Kirkpatrick. 2002. Historical progress of biodiversity conservation in the protected-area system of Tasmania, Australia. *Conservation Biology* 16:1–11.

Myers, N., R. A. Mittermeier, C. G. Mittermeier, G. A. B. da Fonseca, and J. Kent. 2000. Biodiversity hotspots for conservation priorities. *Nature* 403:853–58.

Ogden, J., and G. H. Stewart. 1995. Community dynamics of the New Zealand conifers. Pp. 81–119 in *Ecology of the southern conifers*. Ed. N. J. Enright and R. S. Hill. Carlton: Melbourne Univ. Press.

Olson, D. M., and E. Dinerstein. 1998. The Global 200: A representation approach to conserving the Earth's most biologically valuable ecoregions. *Conservation Biology* 12:502–15.

Podger, F. D., and M. J. Brown. 1989. Vegetation damage caused by *Phytophthora cinnamomi* on disturbed sites in temperate rainforest in western Tasmania. *Australian Journal of Botany* 37:443–80.

Stewart, G. H., and T. T. Veblen. 1982. Regeneration patterns in southern rata (*Metrosideros umbellata*)–kamahi (*Weinmannia racemosa*) forest in central Westland, New Zealand. *New Zealand Journal of Botany* 20:55–72.

Wardle, P. 1991. *Vegetation of New Zealand*. Melbourne: Cambridge Univ. Press.

Webb, L. J. 1978. A structural comparison of New Zealand and southeastern Australian rainforests and their tropical affinities. *Australian Journal of Ecology* 3:7–21.

White, M. E. 1986. *The greening of Gondwana*. Sydney: Reed.

Williams, W. D. 1974. Introduction. Pp. 3–15 in *Biogeography and ecology in Tasmania*. Vol. 25 of *Monographiae Biologicae*. Ed. W. D. Willams. The Hague: Dr. W. Junk Publishers.

CHAPTER 9

Rainforests at the Margins: Regional Profiles

Dominick A. DellaSala

While this book used a global model for delineating the temperate rainforests of the world as presented in Chapter 1, there were some regions that appeared on closer inspection to qualify as rainforest even though they did not make the cut in the global model. The authors of regional profiles in this chapter make a compelling case for inclusion of these areas as "outliers" or "rainforests at the margins." These are regions where local climates—especially the presence of fog—appear to compensate for climatic conditions otherwise atypical of more-definitive rainforests. For instance, in the Colchic and Hyrcanic forests of the Western Eurasian Caucasus, fog provides ample conditions for oceanic lichens and humidity-dependent vegetation despite temperatures considered too warm for most temperate rainforests.

On the other hand, low temperatures, fog, and high humidity combine to compensate for relatively low annual precipitation in the rainforests of the Russian Far East, northeast China, Southern Siberia, and the Eastern Korean Peninsula. Likewise, for the Knysna-Tsitsikamma forests of South Africa, fog is compensatory for relatively low precipitation and allows mesic species typical of temperate rainforests to flourish. As fog was not included in the global rainforest model and yet there was evidence of rainforest communities in these regions, we felt it was necessary to present them as part of the global network of rainforests in order to inspire additional research into how temperate rainforests from distant parts of the globe persist under climatic conditions atypical of rainforests.

REGIONAL PROFILE: COLCHIC AND HYRCANIC TEMPERATE RAINFORESTS OF THE WESTERN EURASIAN CAUCASUS

George Nakhutsrishvili, Nugzar Zazanashvili, and Ketevan Batsatsashvili

Located in the eastern (E, SE, NE) portion of the Black Sea catchment basin (see figure 9-1), the climate of the Colchic region is moderately warm (24–25°C) with cool (4–6°C) winters and abundant annual precipitation (typically 1,800–2,200 millimeters and up to 4,500 millimeters on Mount Mtirala). The Hyrcanic region covers the eastern slopes of the Talysh Mountains and northern slopes of the Elburz Mountains at the southern coastal area of the Caspian Sea. The regional climate is also warm (24–26°C) with colder (–2+6°C) winters and less precipitation (~1,500 millimeters to 2,000 millimeters annually). Collectively, the two regions extend from 35 to 44°N latitude and are part of a much larger region known as the Caucasus, an area dominated by the Caucasus Mountains that mark the boundary between Europe and Asia. Here, a thin band of rich rainforest is surrounded by expansive deserts, mountains, and inland seas.

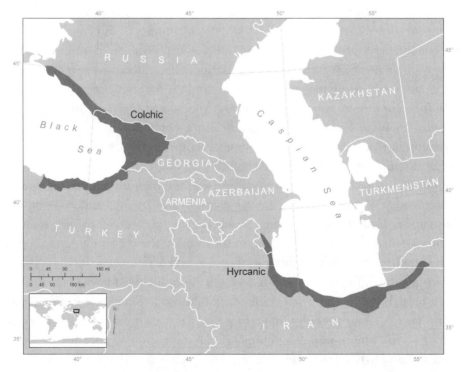

Figure 9-1. Colchic and Hyrcanic temperate rainforests of the Western Eurasian Caucasus.

Colchic straddles two phytogeographical regions, including the Euxinian (or "Hospitable Sea," an ancient Greek name for the Black Sea) province of the Eurosiberian phytogeographical region, and the Hyrcanic (from an ancient Greek name for the region) province of the Irano-Turanian region (Takhtajan 1978). The Colchic and Hyrcanic forests, located within these regions, are the most important relicts of the Arcto-Tertiary forests of Western Eurasia, with many relict and endemic plants and rare fauna (Scharnweber et al. 2007). Many plants have ancient boreal affinities from the Tertiary period, and, therefore, the Caucasus is considered a global "hot spot"—an area where numerous species are highly concentrated—as recognized by Conservation International (also see nationalgeographic.com[1]), and a globally unique ecoregion as recognized by the World Wildlife Fund. The Hyrcanic forests of the Tallish have also been nominated for World Heritage status (see Scharnweber et al. 2007).

A CASE FOR TEMPERATE RAINFOREST

Making a case for the region (both forest types) as warm–temperate rainforest requires closer examination of regional climatic attributes not detectable using the global rainforest model. Here, we describe some of the unique climatic conditions that should help to qualify the Hyrcanic and Colchic forests as a temperate region with rainforest characteristics.

Temperate Climate

Although both regions have temperature levels exceeding levels used in the global temperate rainforest model (e.g., a mean warmest-quarter temperature of 20°C), and some authors consider the Colchic rainforest subtropical (Rikli 1943) but with temperate tree composition (Lavrenko 1958; Dolukhanov 1980), nevertheless this region should qualify as temperate. For instance, Colchic cannot be considered subtropical climatically or structurally in terms of plant communities: air temperatures are lower, seasonal distribution of rainfall is largely continuous, and there is no broadleaf evergreen forest as in subtropical regions.

Hyrcanic is located slightly to the south of Colchic, and therefore the climate is somewhat warmer and in places drier. Unlike the more continuous precipitation of Colchic, Hyrcanic's climate has large amounts of precipitation only at relatively low altitudes. The thermal gradient is approximately the same

---

[1]www.nationalgeographic.com/wildworld/profiles/terrestrial/pa/pa0422.html

as for Colchic: favorable conditions for rainforest are restricted to montane zones, mostly along the Tallish Mountains and the western sector of the Elburz Mountains up to 800–1,000 meters. However, the eastern sector is drier, with significantly reduced numbers of typical Hyrcanic species and an increased portion of Irano-Thuranian species. Therefore, Hyrcanic's climate is closer to temperate than subtropical because of a quite distinct dry period in the second half of spring and summer (Gerasimov 1966). Further, in the explanatory text to the Map of Natural Vegetation of Europe, both the Colchic and Hyrcanic forests are considered humidity- and warmth-requiring (hygro-thermophilous) broadleaf forests (Dolukhanov 1980; Nakhutsrishvili 1999; Doluchanov and Nachucrivšvili 2003).

Moisture-Trapping Mountains

Evidence for temperate rainforest in the Caucasus is generally related to the same principal characteristics that define most of the other regions covered in this book: mountainous chains located along coastlines, which trap a large portion of the moisture arising from oceanic air masses on their windward side. In the Caucasus, these barriers are formed by a topographical triangle created by the intersection of the western part of the Greater Caucasus Mountain Range (Georgia, Russia), the western part of the Lesser Caucasus Mountains (Turkey, Georgia) and Likhi Ridge (the bridge ridge between the Greater and Lesser Caucasus, Georgia) at the Black Sea, and by the Talysh-Elburz Mountain Range along the south-southwestern coast of the Caspian Sea (Iran, Azerbaijan). The warm and humid climate of this region has been present since the late Tertiary, creating the primary reason why the Caucasus has acted as a shelter for hygro-thermophilous relicts during the previous ice age. Consequently, Colchic and Hyrcanic forests are the oldest forests in Western Eurasia in terms of their origin and evolutionary history, the most diverse in terms of relict and endemic woody species and tree diversity, and the most natural in terms of transformation of historic structure.

Both Colchic and Hyrcanic, whose formation is attributed to the Upper Pliocene (Kolakovsky 1961), have a number of common features. However, they also have markedly defined originality as described below.

COLCHIC RAINFORESTS

The total area of all forests of the Colchic region, estimated by the GIS unit of WWF Caucasus using Google images, is ~3 million hectares. There are a number of forest types belonging to this region: lowland hardwood forests; foggy gorges and mixed broadleaf forests; sweet chestnut (*Castanea sativa*) forest; beech (*Fagus orientalis*) forest; dark coniferous forest; and oak (*Quercus* spp.) woodland. But the main distinguishing feature of these forests is the half-prostrate shrubs characterized by vegetative reproduction, forming dense understories up to 4 meters high. Forests are marked by broadleaf evergreens, including several *Rhododendrons* (*R. ponticum*, *R. ungernii*, *R. smirnowii*, the last two being local endemics of southern Colchis), cherry-laurel (*Laurocerasus officinalis*), Black Sea holly (*Ilex colchica*), as well as deciduous mountain cranberry (*Vaccinium arctostaphylos*), and Oriental viburnum (*Viburnum orientale*) (Dolukhanov 1980; Zazanashvili 1999; see plate 14).

Wildlife includes the relict and endemic Caucasian salamander (*Mertensiella caucasica*) and the Caucasian viper (*Vipera darevskii*), which can be found in the Lesser Caucasus. Notably, there is evidence that the Caucasian salamander is a Miocene relict. Several other species are on the IUCN Red List, 2008 (www.redlist.org) as globally threatened (see table 9-1).

HYRCANIC RAINFORESTS

The total estimated area of Hyrcanic rainforests, which covers all types of forests of the Talysh and Elburz Mountains, is around 1.96 million hectares—1.85 million hectares in Iran (Ministry of Ecology and Natural Resources of Azerbaijan and Iranian Cultural Heritage, Handicrafts and Tourism Organization 2009), and ~108,000 hectares in Azerbaijan (Talysh Mountains).[2]

These forests are strongly transformed in the lowlands and foothills of both regions because of severe human intrusions (e.g., see Scharnweber et al. 2007). They include examples of oak and mixed broadleaf forest, ironwood forest, and ravine forest types rich in narrow endemics, including: chestnut-leaved oak (*Quercus castaneifolia*), Persian ironwood (*Parrotia persica*), Caucasian alder (*Alnus subcordata*), silktree (*Albizia julibrissin*), and Caspian locust (*Gleditsia caspia*). Lianas and shrubs such as Pastuchov's ivy (*Hedera pastuchowii*) and Alexandrian laurel (*Danae racemosa*) are especially prominent. Some endemics are taxonomically very similar to relict species of the Colchic refugium, such as Hyrcanian ruscus (*Ruscus hyrcanus*), Hyrcanian winterberry (*Ilex hyrcana*), and Hyrcanian box (*Buxus hyrcana*) (Dolukhanov 1980). There also are common relict species, such Caucasian zelkova (*Zelkova carpinifolia*), wing nut (*Pterocarya pterocarpa*),

---

[2]www.cac-biodiversity.org/aze/aze_forestry.htm

Table 9-1. Globally threatened species and subspecies of the Colchic and Hyrcanic regions of the Western Eurasian Caucasus.

Common Name	Scientific Name	Colchic	Hyrcanic	IUCN Red List (2009)
Mammals				
Persian leopard, also known as the Caucasian leopard or Central Asian leopard	*Panthera pardus* ssp. *saxicolor*; *Panthera pardus* ssp. *ciscaucasica*	−	+	EN
Wild goat	*Capra aegagrus*	+	+	VU
West Caucasian tur	*C. caucasica*	+	−	EN
Birds				
Lesser white-fronted goose	*Anser erythropus*	+	+	VU
Imperial eagle	*Aquila heliaca*	+	−	VU
Greater spotted eagle	*Aquila clanga*	+	+	VU
Red-breasted goose	*Branta ruficollis*	+	+	EN
Lesser kestrel	*Falco naumanni*	−	+	VU
Saker Falcon, Saker	*Falco cherrug*	+	+	EN
Siberian crane	*Grus leucogeranus*	−	+	CR
Marbled duck	*Marmaronetta angustirostris*	+	+	VU
White-headed duck	*Oxyura leucocephala*	+	+	EN
Sociable lapwing	*Vanellus gregarius*	−	+	CR
Reptiles				
Clarks felseneidechse (German)	*Darevskia clarkorum*★	+	−	EN
Charnali lizard	*D. dryada*★	+	−	CR
Large-headed water snake	*Natrix megalocephala*★	+	−	VU
Common tortoise	*Testudo graeca*	+	+	VU
Caucasian viper	*Vipera kaznakovi*★	+	−	EN
Pontic viper	*V. pontica*★	+	−	EN
Amphibians				
Caucasian salamander	*Mertensiella caucasica*★	+	−	VU
Persian mountain salamander	*Paradactylodon persicus*★	−	+	NT

Key:

CR: Critically Endangered; EN: Endangered;VU:Vulnerable.

Plus symbol (+) indicates species distribution/occurrence in the region; minus symbol (−) indicates not distributed/occurring.

★ Temperate rainforest is the major habitat.

date plum (*Diospyros lotus*), and mountain cranberry. While in Colchic several species of *Rhododendron* are present, this genus is absent in Hyrcanic, where the understory is generally sparse. Evergreen species are less characteristic for Hyrcanic forests, both in number and structurally. But the total number of woody species is greater than that in Colchic. Wildlife include brown bear (*Ursus arctos*), lynx (*Lynx lynx*), and the highly endangered leopard (*Panthera pardus saxicolor*; see table 9-1).

THREATS

The low economic status of the local people and the high demand locally for natural resources are the primary root causes threatening rainforests in both places. In addition, high human population densities and thousand-year utilization of forests have contributed to widespread forest declines.

In Colchic, dramatic changes in the landscape were observed during the Soviet period (since 1921), including extensive development of lowlands, drainage of wetlands, peat extraction, expansion of agriculture, particularly tea, citrus, and corn plantations, and growth of human settlements. Collectively, these changes eliminated around 90 percent of the lowland and foothill forests. Plantations, however, were abandoned after the collapse of the Soviet Union and may recover as forests in the future. A number of nonnative or adventives, such as eastern baccharis (*Baccharis halimifolia*), water couch (*Paspalum paspalodes*), and broomsedge bluestem (*Andropogon virginicus*), have occupied lowlands and expanded into rainforest. An exotic tree, the princess tree (*Paulownia imperialis*), has been cultivated, becoming a recent aggressive invader with the potential of replacing native species.

Logging and grazing have also triggered the expansion of weeds and invasive species such as Japanese spiraea (*Spiraea japonica*), and logging-related landslides pose another risk, which along with high annual precipitation has had human consequences. For instance, many villages were destroyed in the last few decades because of logging-related landslides in the Adjara region, Georgia (southern Colchic), forcing thousands of people to migrate from their homelands. Other threats include illegal logging of firewood by local poverty-stricken people in certain areas, and the poaching of wildlife, particularly birds. The situation in the mountains remains much better because of Soviet forestry policy developed in the 1960s for Georgia's montane forests, which were designated as protected forests for landslide and erosion management, water regulation, and recreational purposes.

In Hyrcanic, forest decline has been dramatic over nearly five decades of logging and other uses. For instance, in 1963 the Hyrcanic forests of Iran totaled some 3 million hectares but only roughly 1.8 million hectares now remain, a drop of over 40 percent (~0.9 percent annual deforestation rate). Forest conversion to agriculture (principally cultivated rice) has been quite common.

Most notably, according to recent climatological forecasts, by 2100 average temperatures may rise by 3.5°C and the annual precipitation may drop by 10 percent throughout the Caucasus (Shvangiradze and Beritashvili 2009).

CONSERVATION PRIORITIES

Conservation priorities are described in the Caucasus Ecoregional Conservation Plan prepared by more than 150 experts from all 6 countries of the Caucasus (Williams et al. 2006). Both Colchic and Hyrcanic were represented in Priority Conservation Areas (Williams et al. 2006), and a number of conservation targets and actions proposed.

In order to address the threats described above and to more effectively ensure the long-term conservation of globally important forests throughout the Caucasus, tracts of forests should be set aside in effectively managed "econets," consisting of protected areas and linking corridors. Various types of protection regimes need to be explored and new methods tested to conserve additional forests, for example, by developing special regulations for conserving forests. Policies for improving management in the forestry sector need to be worked out, and model projects for promoting sustainable and community-based forestry need to be implemented. Reforestation is required in severely degraded areas and opportunities exist to restore abandoned plantations to forests.

The Caucasus Ecoregional Conservation Plan includes sections on conservation, management, and restoration along with immediate actions, 10-year targets, and long-term goals. Measures that should be taken in the near future include creating protected areas, improving management in existing reserves, training and capacity building, drafting and adopting legislation on forest management, and carrying out model conservation projects. According to this plan, around 13.5 percent of forests are currently in protected areas (Williams et al. 2006); however, in the longer term (15 years), an additional 10 percent should be conserved (IUCN Categories I-IV), which would bring the total forest protection to nearly a quarter of the region's forests. This general target is suitable for both Colchic and Hyrcanic forests.

The core of a protected-areas network should consist of the few remaining semi-natural forests surrounded by buffer zones of varying human intensity. These areas and their buffer zones need to be effectively protected through cooperative management with local populations, which, in turn, can enhance the economic well-being of local communities along with elevated living standards through innovative job creation programs based on sustainable land use and environmentally responsible industries. The ecotourism sector has been well established, centered around the numerous natural attractions.

LITERATURE CITED

Dolukhanov, A. G. 1980. *Colchic underwood*. Tbilisi, Georgia: Metsniereba. (In Russian.)

Doluchanov, G., and G. Nachucrivšvili. 2003. *Die natürlichen Vegetations formationenEuropas und ihre Untergliederung. Hygrophile thermophytische Laubmischwälder*. Pp. 384–88 in *Karte der natürlichen Vegetation Europas (Map of the Natural Vegetation of Europe), Maßshtab/Scale 1:2,500,000*. Ed. U. Bohn, G. Gollub, and C. Hettwer. Federal Agency for Nature Conservation, Bonn, Germany.

Gerasimov, I. P., ed. 1966. *Caucasus*. Moscow: Nauka. (In Russian.)

IUCN 2009. IUCN Red List of Threatened Species. Version 2009. www.iucnredlist.org (active 9/14/09).

Kolakovsky, A. A. 1961. *The plant world of Colchis*. Moscow: Publishing House of Moscow Univ. (In Russian.)

Lavrenko, E. M. 1958. Concerning position of the forest vegetation of the Caucasus in the system of botanical-geographic division of the Palearctic. *Botany Journal* 43 (9):1237–53. (In Russian.)

Ministry of Ecology and Natural Resources of Azerbaijan and Iranian Cultural Heritage, Handicrafts and Tourism Organization. 2009. Baku, Azerbaijan, and Tehran, Iran.

Nakhutsrishvili, G. 1999. *Vegetation of Georgia*. Camerino, Italy: Braun-Blanquetia.

Rikli, M. 1943. Das Pflanzenkleid der Mittelmeerländer. Zweite Auflage. B. I.. Bern.

Scharnweber, T., M. Rietschel, and M. Manthey. 2007. Degradation stages of the Hyrcanian forests in southern Azerbaijan. Archiv für Naturschutz und Landschaftsforschung 133–56.

Shvangiradze, M., and B. Beritashvili, eds. 2009. The 2nd national message of Georgia to U.N.O. Framework Convention, Tbilisi.

Takhtajan, A. L. 1978. *The floristic regions of the world*. Leningrad: Nauka (In Russian.)

Williams L., N. Zazanashvili, G. Sanadiradze, and A. Kandaurov, eds. 2006. Ecoregional Conservation Plan for the Caucasus. WWF, KFW, BMZ, CEPF, MacArthur Foundation. Printed by Signar Ltd. 220. www.assets.panda.org/downloads/ecp__second _edition.pdf

Zazanashvili, N. 1999. *On the Colkhic vegetation*. Pp. 191–97 in *Conference on recent shifts in vegetation boundaries of deciduous forests, especially due to general global warming*. Ed. F. Klötzli and F. G. R. Walther. Basel-Boston-Berlin: Birkhäuser Verlag.

REGIONAL PROFILE: HUMIDITY-DEPENDENT FORESTS OF THE RUSSIAN FAR EAST, INLAND SOUTHERN SIBERIA, AND THE EASTERN KOREAN PENINSULA

Pavel V. Krestov, Dina I. Nazimova, Nikolai V. Stepanov, and Dominick A. DellaSala

Continental Asia, in the transitional area between the temperate and boreal zones, has three regions with relatively high humidity compared to its surroundings that may qualify forests in the region as rainforests despite most of the region having temperatures too low to qualify as definitive rainforests: (1) the Pacific Coast of the Russian Far East (43–52°N latitude); (2) mountainous areas of Inland Southern Siberia (53–56°N latitude); and (3) the Eastern Korean Peninsula (36–42°N latitude; see figure 9-2; note that the Eastern Korean Peninsula was included in the rainforest distribution model but is discussed here briefly based on author expertise). Despite mean annual temperatures in much of the region ranging from −2 to 4°C, rainforests there resemble those of the Pacific Northwest structurally and include many relict and endemic plants and animals that require humid conditions with no periods of water deficit. These ecosystems could qualify as humidity-dependent forests or "rainforests"

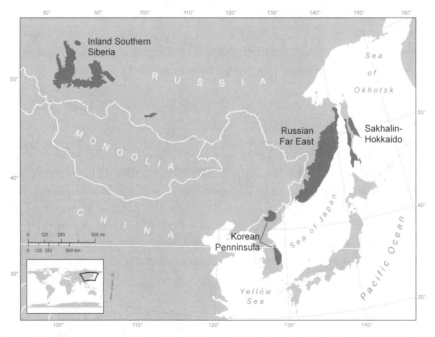

Figure 9-2. Temperate and boreal rainforest of the Russian Far East, Inland Southern Siberia, and the Eastern Korean Peninsula.

using the terminology in this book. They occur in transition between cool-temperate and boreal.

Within these regions, relict forests consist of ancient taxa that survived drastic climatic shifts in the Pleistocene because they were protected by localized climatic refugia provided by mountainous regions and mild perhumid climates. Permanent high humidity levels during the Holocene historically maintained humidity-dependent species in typical temperate rainforests (Grichuk 1984). Relatively deficient precipitation levels, compared to those of classic temperate rainforests, today are compensated by reduced evapotranspiration rates because of low temperatures. Thus, plants receive moisture levels similar to those of definitive temperate rainforests.

THE RUSSIAN FAR EAST

An estimated 6.8 million hectares of cool-temperate rainforest, also known as "Ussuri taiga," occurred historically in this region, with roughly 40 percent (2.8 million hectares) remaining (Koryakin 2007). Of this, approximately 1.5 million hectares of the core area of Ussuri taiga in the central Sikhote-Alin Mountains were recognized by UNESCO as a World Heritage site (whc.unesco .org/en/list/766). Similar rainforests in northeast China; however, were totally devastated by centuries of human influence. Therefore, most of the discussion here will focus on what remains in the Russian Far East.

The southern mountain system of the Russian Far East, Sikhote-Alin, is strongly affected by monsoon climate. Therefore, the cold period here is drier than the warm period, with shallow snow (30–50 centimeters). Most precipitation falls in warm months. During the growing period, plants receive from 800 to 1,000 millimeters of rainfall. Contrasting temperatures between land and sea cause the formation of fog, which reduces ambient temperature and prevents intensive evaporation from the ground. A lack of moisture deficits, in combination with moderate heat (growth-degree-day index greater than 10°C) and long growing seasons, provide ideal conditions for species-rich and humidity-dependent forests to flourish (Krestov 2003). Here, forests consist of Jezo spruce (*Picea jezoensis*) and Manchurian fir (*Abies nephrolepis*) at elevations from 700–800 to 1,200–1,400 meters (orotemperate belt), with needle fir (*Abies holophylla*), Korean pine (*Pinus koraiensis*), and several hardwoods (*Fraxinus* spp., *Tilia* spp., *Phellodendron amurensis, Juglans* spp.) at lower elevations (Nakamura and Krestov 2005, plate 15 ab). The broadleaf Korean pine component is a zonal vegetation type located in the sub-maritime sector of the northern temperate subzone of the Sikhote-Alin Mountains (southernmost part of the Russian Far East) and historically in northeast China.

Cooler summers and milder winters characteristic of a monsoon climate provide favorable conditions for frost intolerant plants (e.g., *Abelia koreana*, *Aralia continentalis*, *Bergenia pacifica*, *Ilex rugosa*, *Philadelphus tenuifolius*), which reach their northernmost limits in these forests. Smaller differences between summer and winter temperatures, in general, are characteristic of oceanic regions, and these conditions support the development of humidity-dependent forests as well as provide climatic refugia during times of climate change. The endemic genus *Microbiota*, which occurs along the Sikhote-Alin Mountains, and species with a Japanese distribution that are isolated on the coast of Asian mainland, including Fujiyama (Fauriei's) rhododendron (*Rhododendron brachycarpum*), oval-leaf blueberry (*Vaccinium ovalifolium*), redberry (*V. praestans*), and Japanese primrose (*Primula jezoana*), are all good examples of humidity-dependent species that persisted during the Pleistocene dry spells. Because of a long history (since the Tertiary) of uninterrupted forest development, closeness to the species-rich East Asian floristic center (Takhtajan 1986) and high humidity, native plants occupy all possible niches in forest ecosystems.

Despite a high diversity of canopy species, the most important processes for wildlife in these forests are controlled by Korean pine, a moderately shade-tolerant species that lives up to 500–600 years (Nakamura and Krestov 2005). Seeds of the stone pine are an important source of energy for well-developed food chains supporting small rodents (squirrel—*Sciurus vulgaris mantchuricus*; chipmunk—*Tamias sibiricus*), and birds (nutcrackers—*Nucifraga caryocatactes*; nuthatches—*Sitta europaea*) specializing on pine seeds, as well as Far Eastern megafauna such as bears (*Ursus arctos beringianus*, *U. thibetanus*) and wild boars (*Sus scrofa*). These species follow seed crops of stone pine and Mongolian oak (*Quercus mongolica*) and, in turn, support numerous predators.

On Sakhalin and on the southern isles of the Kuril arc (part of the Pacific Ring of Fire—the Pacific volcanic zone), Sakhalin fir (*A. sachalinensis*) forms pure stands in conditions of cool summers, mild winters, and precipitation greater than 1,500 millimeters, with half or more of that falling in winter. Humid fir forests of the Far East are characterized by presence of many representatives of Japanese flora, including humidity-dependent species that survived severe Pleistocene cooling and aridization due to refugia capable of holding moisture. The species composition of Sakhalin forests is enriched by representatives of the Japanese flora, including rugose holly (*Ilex rugosa*), creeping skimmia (*Skimmia repens*), evergreen huckleberry (*Vaccinium ovatum*), and pubescent huckleberry (*V. hirtum*).

The rather wide distribution of these forest types from coast to inland, where there is a sharp transition from an oceanic monsoon to a continental climate, has led to marked heterogeneity of vegetation types. The forests on the eastern slopes of the Sikhote-Alin Mountains facing the Sea of Japan receive enough moisture (~1,000 millimeters annually) to approximate definitive rainforest conditions. Even though the mountaintops hardly reach 2,000 meters elevation, the Sikhote-Alin range traps air masses from the Sea of Japan and deposits them on the western slopes, where rainfall in months with positive temperature can reach 1,100 millimeters. Fog originating from differences in air and sea temperatures, and the coincidence of summer temperature and precipitation, appears to compensate for comparatively low annual precipitation, supporting development of tree-dwelling mosses and lichens reminiscent of rainforests of the Pacific Northwest (Goward and Spribille 2005). As in the rainforests of Inland Northwestern North America (see chapter 3), belts of humid forests also are found on the windward (eastern) sides of mountain ranges farther inland, which capture the high orographic precipitation associated with the Badzhal'skiy, Bureinskiy, and Lesser Xingan Mountains (Vitvitsky 1961).

The humid environment in these forests reduces the hazard of fires during most of the summer; however, in fall, when the Siberian anticyclone increases its influence, the rainfall in Sikhote-Alin and Sakhalin drastically decreases. This, in combination with the accumulation of litter in broadleaf and coniferous forests, creates conditions for fires, which are initiated mostly by people and sometimes by dry lightening. In the interior areas of the Russian Far East, the probability of spring fires increases due to a greater spring moisture deficit caused by quick melting and evaporation of shallow snow.

Sikhote-Alin forests have served as a species-rich refuge during the Pleistocene ice age that now includes over 100 globally unique (endemic) plant and animal species with origins dating back to the Tertiary boreo-tropical (Arcto-Tertiary) biome (the time between the extinction of the dinosaurs and the beginning of the Pleistocene ice age). Exemplary endemics include the world's largest cat, the Amur tiger (*Panthera tigris atlaica*), an isolated and very small population of snow leopard (*Panthera pardus orientalis*), and the world's last remaining wild population of ginseng (*Panax ginseng*). Thus, the composition of these forests can be traced back to ancestral boreo-tropical flora that occupied very extensive areas in the temperate and polar latitudes of the Northern Hemisphere during the Tertiary. Humidity-dependent vegetation survived in North Asia because it was unaffected by glaciers during the Pleistocene, or indeed since the high temperatures reached during the Pliocene optimum (Grichuk 1984). Most of the present-day plants have ancestral taxa dating to Tertiary floras. In the most severe Pleistocene glaciations of 18–20,000 years ago, Jezo spruce kept its dominant status in the lower montane belt of the Sikhote-Alin range, which possibly was a refugium for a number of temperate species. Within the montane conifer forest belt of the Sikhote-Alin Mountains several species existed in isolated localities far from their main range in insular Asia.

INLAND SOUTHERN SIBERIA

This interior region historically contained about 7 million hectares of montane and humidity-dependent forests (Polikarpov et al. 1986), most of which (6 million hectares) is still present in the Sayany, Altai, and Hamar-Daban (just south of Lake Baikal) Mountains. Of this, approximately 1.6 million hectares in the Altai Mountains includes large tracts of humidity-dependent forests that were recognized by UNESCO as a World Heritage site (whc.unesco.org/en/list/768). Forests are wet enough to qualify as humidity-dependent in the hemiboreal (halfway between temperate and boreal) and southern boreal zones of southern Siberia, where they are dominated principally by Siberian fir (*A. sibirica*) in lower elevations and Siberian pine (*P. sibirica*) in higher forest belts (Nazimova and Polikarpov 1996; plate 15b). Like their Far Eastern counterparts, these forests represent relict vegetation of modern refugia that occur elsewhere in drier continental regions of Northern Asia.

Inland Southern Siberia is a place where zonal vegetation of the Siberian plains, represented by steppe, shifts in the mountains of Altai, Sayany, and Hamar-Daban to humid and perhumid forests dominated by broadleaf trees such as silver birch (*Betula pendula*) and Eurasian aspen (*Populus tremula*), and conifers such as Scots pine (*P. sylvestris*), Siberian fir, and Siberian pine. No one would expect humid forests here due to the semiarid surroundings. However, the mountainous systems with snowy tops in the summer act as a barrier to Atlantic air masses, which dump more than 1,500 millimeters of precipitation annually, an amount that is nearly evenly distributed seasonally. The northern macroslope of the Hamar-Daban Mountains is strongly affected by moisture condensed by the proximity of very cold air from Lake Baikal and warm and damp Atlantic air masses. The relatively high heat supply in the summer (growth-degree-day index greater than10°C) and humidity in isolated pockets have led to the formation of rainforests characterized by boreal dominants with well-developed understories, most of which are temperate (nemoral). Old Siberian firs up to 150–180 years old and Siberian pines up to 600 years old grow to 40 meters tall and occupy windward slopes, acting as climate refugia for relict humidity-dependent species.

Notably, Siberian fir forms pure stands in the middle elevations in conditions of high humidity, with annual rainfall exceeding 1,000 millimeters. In the lower part of the belt, fir mixes with aspen and Siberian pine on well-drained microsites. The tree layer is very simple and usually includes two stories represented by the same species in different ages. Most of temperate (nemoral) species occupying these forests (*Actaea spicata, Asarum europaeum, Brachypodium sylvaticum, Brunnera sibirica, Carex sylvatica, Chrysosplenium ovalifolium, Daphne mesereum, Dentaria sibirica, Festuca altissima, Frangula alnus, Sanicula europaea*, and *Stachys sylvatica*) belong to European and/or Mediterranean flora at the eastern edge of their distributions. However, some Tertiary relicts (*Anemone baicalense, Menispermum dahuricum*, and *Waldsteinia ternata*) are of East Asian origin. Of more than 800 species of lichens inhabiting the forests of Altai and Sayany (Sedel'nikova 2001), over 120 species are at the northern limits of their ranges that expand far southward to the subtropical zone. Among them, there are specific humidity-indicating lichens (*Sticta limbata, S. wrightii, S. fuliginosa*, and *S. sylvatica*), which, growing on tree trunks, are considered indicators of definitive temperate and boreal rainforests all over the world (Goward and Spribille 2005).

The ancestral vegetation of these forest types was heavily suppressed by climatic shifts during the Pleistocene and, consequently, they could not maintain composition or structure until the present climate developed. However, the mountains of southern Siberia are climatic refugia for humidity-dependent epiphytic cryptogams ("lower plants" that reproduce by spores), forbs, and shrubs that probably constituted different community types before the Pleistocene coolings and aridizations. Perhaps ecosystems of this type served in the Pleistocene as refuge for the Siberian lime tree (*Tilia sibirica*), which at present occurs in warmer and drier climates. The modern structure and species composition of Southern Siberia humidity-dependent forests include a rich mixture of plants characteristic of subalpine meadows, temperate (nemoral) forests, and boreal forests. The modern humid and mild climate corresponds to northern temperate forests in general, but these forests keep their "boreal" appearance due to long isolation from the distribution areas of potential nemoral dominants.

THE EASTERN KOREAN PENINSULA

While there are no data on the historic extent of these cool-temperate rainforests, today about 1.87 million hectares are present in the high mountains along the eastern coast of the Korean peninsula (Taebaek mountain systems) between 36 and 42°N latitude. The region receives up to 1,200 millimeters of precipitation annually, of which about half falls in the winter, with annual temperature varying from 4 to 8°C from mountaintops to lowlands, respectively. Mt. Sulak (1,708 meters elevation), Taebaek (1,561 meters elevation), and Odae (1,563 meters elevation) have well-expressed vertical zonation spanning temperate mixed coniferous and broadleaf deciduous forests. These montane areas include a broad array of forest types, from Mongolian oak (*Quercus mongolica*) and Korean pine (*P. koraiensis*), to coniferous forests dominated by spruce (*P. jezoensis*) and fir (*A. nephrolepis*), to subalpine krummholz dominated by Siberian dwarf pine (*P. pumila*) reaching the southernmost part of its distribution (Song 1991; 1992). Such forests physiognomically are very close to Sikhote-Alin forests; however, high precipitation and relatively high winter temperatures favor humidity-dependent species such as Siebold's ash (*Fraxinus sieboldii*), Japanese spicebush (*Lindera obtusiloba*), Oyama magnolia (*Magnolia sieboldii*), mountain sumac (*Rhus trichocarpa*), dwarf bamboo (*Sasamorpha borealis*), fragrant snowbell (*Styrax obassia*), and others.

THREATS

The humidity-dependent boreal and cool-temperate rainforests of Northern Asia represent relict vegetation complexes formed and widely distributed during the Tertiary, millions of years ago. Great aridization in the Pleistocene epoch swept this vegetation type from nearly the entire continent (Grichuk, 1984). The Holocene humidization revived remnants of these relict ecosystems, which are controlled by current climatic conditions, especially high humidity levels. While the history of climate change in most cases explains the rarity of these forests in continental Asia today, the most important contemporary disturbance factors are logging and human-initiated fires. Against the background of anthropogenically transformed areas, such natural disturbances as winds, natural fires, and forest declines by insects pale in comparison, although this could change soon with the advent of global climate change (see below).

A century of industrial forestry has reduced the area of virgin humidity-dependent forests of North Asia drastically. The coniferous forests of Siberian and Korean pines and Jezo spruce suffered more than others because of the high value of timber. Despite big losses, the Russian part of Asia stills has extensive humidity-dependent forests, some of them at different stages of post-logging and post-fire recovery.

Major humidity-dependent cool-temperate rainforests in the Russian Far

East have decreased considerably. According to forest records from the 1890s, the broadleaf Korean pine forests on Sikhote-Alin covered an area of about 68,000 square kilometers (6.8 million hectares). This has declined by about 40,000 square kilometers (4 million hectares—Koryakin 2007). Approximately 21 percent of this area, mainly in barely accessible places, is today represented by intact broadleaf Korean pine rainforests (Aksenov et al. 2006). Most of the rainforests of the Russian Far East and Southern Siberia were logged in the 1960s and 1970s, except Sakhalin Island, the southern part of which was completely logged prior to 1945.

Predominantly selective logging was applied in the forests of the Russian Far East. From forest canopies dominated by 5–6 dominant species, only commercially valuable Korean pine trees were removed. Therefore, even after logging forests retained their forest environment and most of their original species composition. If logging is not followed by human-initiated fires, Korean pine has a chance to be reestablish itself from saplings remaining after logging or by animals specializing on Korean pine seeds, especially squirrels, chipmunks, nutcrackers, and nuthatches. Regeneration of Korean pine, even in favorable conditions, takes a long time (~150 years), the period necessary for pines to reach a canopy. Such secondary forests dominated by broadleaf species with regenerating Korean pine represent nearly half of the logged rainforests in the Sikhote-Alin Mountains, about 18,000 square kilometers (1.8 million hectares; Koryakin 2007). The remainder of logged stands experienced much deeper transformations by fires or repeated loggings, or they had heavy disturbances of forest soils, which have shifted species composition and ecosystem dynamics. These forests need to be excluded from regional economic activities and greater investments must be made in their propagation.

On Sakhalin Island, large tracts of intact humidity-dependent boreal forests remained only in northern half of the island after World War II. Before the 1960s, forests were harvested for local demands only. Industrial logging started in the 1950s and has continued at a rate of about 3.5–4.0 million cubic meters of timber annually. This situation for rainforest conditions has been made even worse by very extensive forest fires on Sakhalin, the largest of which were in 1949 (353,900 hectares), 1950 (207,000 hectares), 1954 (434,600 hectares), and 1989 (217,100 hectares).[3] Another important threat for Sakhalin forests is the global natural gas and oil industry (Sakhalin-I and Sakhalin-II)[4] that in the last decades have constructed a series of pipelines that crisscross the island, totaling

---

[3]www.forest.ru/rus/bulletin/10/2.html

[4]www.en.wikipedia.org/wiki/Sakhalin-II

some 1,300 kilometers. At present only 2.1 million hectares of humid fir and spruce forests remain in more remote parts of the island. None of them are protected in nature reserves.

In humidity-dependent boreal rainforests of Inland Southern Siberia, industrial logging has led to the near elimination of significant populations of Siberian pine because of its valuable timber. Regeneration of pine takes a very long time due to the inhibiting effect of the tall forb cover in these ecosystems. The population of Siberian fir has declined less, and extensive tracts of this forest type can be found across the range of isolated Siberian rainforests. They form a unique ecosystem that is very sensitive to the equilibrium between air humidity and changing temperature. Because both species form the monodominant stands, clear-cutting has been the most common method of logging. This kind of disturbance in the conditions of Southern Siberia inevitably leads to a critical change of species composition and to ecosystem changes from a multilayered forest stand with sparse forb cover to a stand with a single or no tree layer and very dense forbs that exclude regeneration of pine and fir for several decades. Just under 20 percent of the primeval forests of this type are protected in 3 nature reserves (the Ussuriysky and Botchinsky nature reserves, and the Sikhote-Alinskii biosphere) and in the areas with different conservation status.

Generally, the most effective means of ecosystem conservation in all three humidity-dependent regions is through their protection in nature reserves (Center for Russian nature conservation, www.wild-russia.org; Petrosyan et al. 2003–2006), that is, territories with restricted admittance and fully prohibited anthropogenic activities. In the Russian Far East, only 1,200 hectares of the intact humidity-dependent forests are protected in nature reserves, and about 770,000 hectares are conserved at lower protection levels (i.e., "water-protection forests" along rivers, "slope-protection forests," and other designations). In addition, 2,000 hectares of intact humidity-dependent boreal rainforests are in nature reserves in Southern Siberia despite approximately 4 million hectares or two-thirds of their total area still intact; and in Korea there are no intact areas or nature reserves with these forests.

Notably, humidity-dependent forests are effectively preserved in four nature reserves in the Russian Far East: Sikhote-Alinsky (central Sikhote-Alin Mountains),[5] Ussuriysky (southern Sikhote-Alin Mountains),[6] Botchinsky (northern Sikhote-Alin Mountains), and Lazovsky (southern Sikhote-Alin

---

[5]www.wild-russia.org/bioregion13/sikhote/13_sikhote.htm
[6]www.wild-russia.org/bioregion13/13_USSURISKY/13_ussur.htm

Mountains).[7] Four additional nature reserves with similar forests occur in southern Siberia: Altaysky (Altai),[8] Katunsky (Altai),[9] Ubsunurskaya Kotlovina (Tuva Republic), Kuznetskiy Alatau (Kuznetskiy Alatau Mountains)[10] and Baikalsky (Hamar-Daban Mountains).

Overall protection in nature reserves is far too low to conserve humidity-dependent forests. There is an urgent need to connect these areas via wildlife corridors. The major efforts must be undertaken initially in intact or slightly transformed humidity-dependent forests currently outside of nature reserves. Several important steps have been taken by government, beginning in the 1990s. For instance, harvest was precluded in Korean pine forests in the Russian Far East, some existing nature reserves were extended (Sikhote-Alinsky Nature Reserve), and new nature reserves were created (Bastak, Botchinsky, Katunsky) in territories with humidity-dependent forests that maintain their regeneration potential although previously logged. After the collapse of the Soviet Union, the major problems in forest conservation have been a government that has lost control of local logging companies and growing demands for timber in Russia and China, which create additional pressures. Illegal logging also drastically increased beginning in the 1990s as result of protracted reorganization of the forest-management system.

Fortunately for these forests, ongoing worldwide conservation programs to save rare animals also include protection of their habitats. The Ussuri tiger and snow leopard are at the top of food chains within the Korean stone pine ecosystem and appear to be most influenced by forest loss due to the reduction both of feeding territory and of a number of key prey species. Only ~450 tigers and 29 leopards remain in the Sikhote-Alin Mountains, numbers believed too low to sustain viable populations. Strict protection by numerous conservation programs has slowed declines of the humidity-dependent forests used by these animals. The major task now is to avoid fragmentation of habitats, including protection of intact Ussuri taiga for these and other rainforest species. Also, there is a need for baseline data on these forests to know if there are unique vascular plants, lichens, moss, and fungi, particularly those that may be very sensitive to future climatic changes and ongoing fragmentation and deforestation.

The effects of climate change have also become noticeable in the recent decades with an increase of 1°C in the eastern part of Asia (Japan Meteorolog-

[7]www.wild-russia.org/bioregion13/lazovsky/lazovsky.htm

[8]www.wild-russia.org/bioregion9/9-altai.htm

[9]www.wild-russia.org/bioregion9/9-katunski.htm

[10]www.wild-russia.org/bioregion9/9-kuznetsky.htm

ical Agency 2009). In far-inland continental areas such as Southern Siberia, such temperature increases trigger decreases in precipitation rates, which may lead in turn to the decline or even disappearance of humidity-dependent forests. Changes in coastal areas will not likely affect forest ecosystems as much, because of anticipated corresponding increases in precipitation. However, the following changes in temperate rainforest are generally anticipated: 1) relict vegetation that survived the Pleistocene maximum and endemic species on the mountain summits may disappear; 2) an expansion of parasites may affect animal populations; and 3) a rapid expansion of areas with arid climates may destroy humidity-dependent ecosystems (Soja et al. 2007).

Clearly, these are trying times for the humidity-dependent forests of all three regions, which face challenges not unlike those confronting temperate and boreal rainforests around the world. Bold conservation steps are needed soon if these forests are going to hold onto their rightful place among the world's temperate rainforests.

LITERATURE CITED

Aksenov, D. E., M. Y. Dubinin, M. L. Karpachevskiy, N. S. Liksakova, V. E. Skvortsov, D. Y. Smirnov, and T. O. Yanitskaya. 2006. Mapping High Conservation Value Forests of Primorsky Kray, Russian Far East. http://www.globalforestwatch.org/english/russia/pdf/HCVF.pdf

Goward, T., and T. Spribille. 2005. Lichenological evidence for the recognition of inland rainforests in western North America. *Journal of Biogeography* 32:1209–19.

Grichuk, V. P. 1984. Late Pleistocene vegetation history. Pp. 155–78 in *Late Quaternary environments of the Soviet Union*. Minneapolis, MN: Univ. of Minnesota Press.

Japan Meteorological Agency. 2009. Climate change monitoring report—2008. www.ds .data.jma.go.jp/tcc/tcc/products/gwp/CCMR2008.pdf

Koryakin, V. N. 2007. *Korean pine-broadleaf forests of the Russian Far East*. Khabarovsk: Far Eastern Institute of Forest Management Press. (In Russian).

Krestov, P. V. 2003. Forest vegetation of easternmost Russia (Russian Far East). Pp. 91–179 in *Forest vegetation of northeast Asia*. Ed. J. Kolbek, M. Šrůtek, and E. O. Box. Dordrecht, Boston, London: Kluwer Academic Publishers.

Nakamura, Y., and P. V. Krestov. 2005. Coniferous forests of the temperate zone of Asia. Pp. 165–220 in *Coniferous forests*. Vol. 6 of *Ecosystems of the World*. Ed. F. A. Andersson. New York, Paris, London, Brussels: Elsevier Academic Press.

Nazimova, D. I., and N. P. Polikarpov. 1996. Forest zones of Siberia as determined by climatic zones and their possible transformations under global change. *Sylva Fennica* 30 (2–3):201–8. (In Russian.)

Petrosyan V. G., U. S. Reshetnikov, S. L. Kuzmin, T. M. Korneeva, U. D. Nuhimovskaya, A. A.

Varshavskiy, A. V. Pavlov, et al. 2003–2006. Information retrieval system for fauna and flora in protected natural areas of the Russian Federation. www.sevin.ru/natreserves/

Polikarpov, N. P., N. M. Tchebakova, and D. I. Nazimova. 1986. *Climate and mountain forests of southern Siberia*. Novosibirsk: Nauka. (In Russian.)

Sedel'nikova, N. V. 2001. *Lichens of western and eastern Sayan*. Novosibirsk: Izdatel'stvo SO RAN. (In Russian.)

Soja, A. J., N. M. Tchebakova, N. H. F. French, M. D. Flannigand, H. H. Shugart, B. J. Stocks, A. I. Sukhinin, E. I. Parfenova, F. S. Chapin III, and P. W. Stackhouse Jr. 2007. Climate-induced boreal forest change: Predictions versus current observations. *Global and Planetary Change* 56 (3–4):274–96.

Song, J. S. 1991. Phytosociology of subalpine coniferous forests in Korea: I. Syntaxonomical interpretation. *Ecological Restoration* 6:1–19.

———. 1992. A comparative phytosociological study of the subalpine coniferous forests in northeastern Asia. *Vegetatio* 98:175–86.

Takhtajan, A. L. 1986. *Floristic regions of the world*. Berkeley, Los Angeles: Univ. of California Press.

Vitvitsky, G. N. 1961. Climate. Pp. 93–115 in *The Far East*. Ed. G. D. Richter. Moscow: Izdatel'stvo AN SSSR. (In Russian.)

REGIONAL PROFILE: KNYSNA–TSITSIKAMMA TEMPERATE RAINFORESTS OF SOUTH AFRICA

Paul E. Hosten and Jeannine M. Rossa

South Africa is renowned for its spectacular plant and wildlife communities, most notably, its famous "fynbos" (fine-leaved sclerophyllous chaparral). It is less well known for its forests, which cover less than one percent (~570 square-kilometers) of South Africa's total land area and yet contribute disproportionately to its biological diversity. We propose that the Knysna–Tsitsikamma Forest of the Southern Cape (see figure 9-3) should be part of the global network of temperate and boreal rainforests described throughout this book. Below, we briefly describe the forest, its ecology, threats, and ongoing conservation efforts.

Figure. 9-3. Knysna–Tsitsikamma temperate rainforest of South Africa (modified from Cowling et al. 1997).

LOCATION AND DESCRIPTION

At the southern tip of Africa lies the Cape Floral Kingdom, only 90,000 square kilometers in size, but having the greatest concentration of vascular plant species in the world—an astounding 9,030 at last count (Goldblatt and Manning 2002). Floral kingdoms are a botanical classification system dividing the world into geographic regions of plant endemism. Conservation International, a U.S. nonprofit organization dedicated to protecting biodiversity, considers the Cape Floral Kingdom to be a global "biodiversity hotspot."

Well-known plant biomes in the Cape Floral Kingdom include the famous "fynbos" (Ericaceous shrublands; pronounced "fayn-bo-s") as well as renoster veld, succulent karoo (desert ecoregion), and subtropical thicket (Cowling et al. 1997). A lesser-known but equally important plant biome within the Kingdom is the Knysna-Tsitsikamma temperate rainforest, which is scattered in patches of varying size along a narrow coastal belt roughly 15–30 kilometers wide by 160 kilometers long, located between the Western Cape town of George and the Eastern Cape's Cape St. Francis (22° 30' – 24° 30' E longitude and 34°S latitude) (see figure 9.3). Although small by worldwide standards—forest patches total only about 235,483 hectares—the Knysna-Tsitsikamma Forest is the largest indigenous forest in South Africa.

The Knysna-Tsitsikamma Forest is naturally patchy due to the steeply-dissected landscape, moisture availability, and hot, dry föhn-type winds ("bergwinds" which blow from the dry interior to the coast) that also control wildfire patterns. A precipitous coastal scarp levels off to a bench or "platform" along the foothills of the steep Outeniqua Mountains (Western Cape of South Africa). Rivers form deep, narrow incisions through the landscape; estuaries have formed at some river mouths. Forest patches flourish in all parts of the landscape: from ~1,220 meters to sea level (Geldenhuys 1993). Historically, fynbos shrubland was the most common plant community adjacent to forest patches, but commercial forest plantations have replaced fynbos in many places.

Dominant tree species include white milkwood[11] (*Sideroxylon inerme*) in the coastal scarp forest; true yellowwood (*Podocarpus latifolius*)—which can grow to 50 meters—in the platform forest along the bench (see plate 16); and black stinkwood or Cape laurel (*Ocotea bullata*) on the mountain slopes. Some trees are 400–800 years old. Forest structure is multilayered and, in many places, classically "impenetrable." In the wettest areas, lianas and epiphytes drape lichen-

---

[11]We present tree and animal communities names in English to aid the reader; however, note that each species has other names among the 11 official languages of South Africa.

covered trees. Coastal scarp soils are nutrient-rich and basic while platform and mountain-slope soils are nutrient-poor and acidic (von Maltitz et al. 2002). Rivers and creeks glow deep orange from high tannin concentrations.

MAKING A CASE FOR TEMPERATE RAINFOREST

Over the decades, the Knysna-Tsitsikamma Forest has been classified as "subtropical rainforest," "afrotemperate forest," or "afromontane forest," depending on the botanist, the scale of analysis, and the criteria used (see van Daalen 1981; Midgley et al. 1997; von Maltitz et al. 2002). Such classifications often lump the Knysna-Tsitsikamma Forest with South African forests to the east or northeast that experience very different temperature and rainfall regimes. Here, we make the case for their inclusion as "warm"-temperate rainforests, following the terminology in Chapter 1.

The climate of the Southern Cape can be classified as "temperate." Mean daily maximum temperatures range from 18 (winter) to 24°C (summer); mean daily minimums range from 9 to 20°C, respectively (Geldenhuys 1993). Frost is infrequent to absent, unlike the forests of South Africa's Drakensberg Mountains, which experience severe frost and freezing temperatures every winter and therefore do not meet the usual definition of "temperate."

The Knysna-Tsitsikamma Forest receives rainfall year-round, unlike the dry forests of the Western Cape (summer drought) or some of the subtropical forests of the Eastern Cape (winter drought). Measured annual precipitation varies from 1,100 millimeters to 1,500 millimeters (Geldenhuys 1994). Although these numbers are below the low end of the rainfall scale for classification as rainforest, like the coastal redwoods (*Sequoia sempivirens*) in the Pacific Northwest (see chapter 2), oceanic fog appears to contribute even more moisture, especially in areas thick with vascular epiphytes and lichens (Geldenhuys 1994). Within the Knysna-Tsitsikamma area, fog contributions have not been well measured. Some data exist from the Mediterranean climate zone located 500 kilometers west. In the Jonkershoek forest near Stellenbosch, orographically-induced fog contributed more than 600 millimeters moisture annually (Schulze 1997). At the top of Cape Town's Table Mountain, oceanic fog provides three times more moisture than the annual rainfall (greater than 5000 millimeters).

The global model for temperate rainforests (see chapter 1) does not include a variable for compensatory fog, which may be why the model does not predict that South Africa should contain temperate rainforests. Moreover, the omission may be due primarily to scale: the small area covered by these forests and the influence of the coastal mountains may be beyond the model's predictive capability.

RAINFOREST ECOLOGY

Forest ecosystems of the Knysna-Tsitsikamma have been predominately shaped by ancient forces that separated Gondwana, resulting in a rich assortment of endemic species with corresponding ancient lineages. Like the Valdivian and Tasmanian rainforests (see chapters 5 and 8), these forests are remnants of a more extensive forest from the warmer Cretaceous Period (145–70 million years ago), when the supercontinent Gondwana was slowly splitting into the continents we know today (Cowling 1992). The persistence of ancient lineages (e.g. *Podocarpus* spp.) provides evidence of this (Schulze 1997). Throughout subsequent climatic changes, the extent of forests across the landscape waxed and waned, reaching a migratory dead end at the southern tip of the continent, while interior deserts further separated the southern forests from those in central Africa (Scott et al. 1997).

As a consequence of this ancient lineage, the Knysna-Tsitsikamma Forest has a rich flora, including the largest herbaceous flora of all South African forests (Geldenhuys 1993), 465 species at last count (von Maltitz et al. 2002). The forest complex covers a range of sites and provides refuge for species from different climatic periods. Sub-canopy trees, ferns, and epiphytes have a higher species-to-family ratio than the other typical forest-growth forms (Geldenhuys 1993).

Mountain forest patches are less structurally complex and species-rich than coastal scarp and platform forests, the result of more frequent burning from bergwind-mediated wildfires (Geldenhuys 1994). Wildfire also controls the extent of forest patches because—unlike fynbos—most forest tree and shrub species are not adapted to regeneration after fire. Within coastal scarp and platform forests, disturbance rates are low, and most trees (70 percent) die standing (Midgley et al. 1997). Therefore, gaps are small and most tree species are shade-tolerant.

Although not limited in distribution to just the forest biome, examples of Southern Cape endemics found in the Knysna-Tsitsikamma Forest include the white bird-of-paradise (*Strelitzia alba*), coal wood (*Lachnostylis hirta*), and the rare rock "lily" (*Gladiolus sempervirens*). Most South African endemics are members of relatively young lineages with numerous close relatives (i.e., found in the fynbos or other post-Pleistocene plant communities—Geldenhuys 1993).

Most of the tree species also have broader distributions than just this temperate rainforest, which is to be expected given their ancient lineages.

Forest amphibians show the opposite pattern: species richness is low, but endemism is high (von Maltitz et al. 2002). Endemic amphibians include the Knysna spiny reed frog (*Afrixalus knysnae*), which is associated with arum lily (*Zantedeschia* spp.), royal ghost frog (*Heleophryne regis*), which is associated with streams, and plain rain frog (*Breviceps fuscus*).

KEYSTONE WILDLIFE

Primates are keystone wildlife species in the Knysna–Tsitsikamma Forest because of their role as seed dispersers. Chacma baboons (*Papio ursinus*) and vervet monkeys (*Cercopithecus aethiops*) consume and disperse seeds and fruit of numerous tree species. Elephants (*Loxodonta africana africana*) were once important seed transporters and kept a network of migratory paths open, affecting forest heterogeneity and forest composition (Graham et al. 2008). Sadly, population pressure and conversion of surrounding areas to agriculture precipitated the end of the Knysna elephant herds. Only a single individual—possibly three—remains, elusive and old. Other keystone species, Cape buffalo (*Syncerus caffer caffer*) and the Cape lion (*Panthera leo melanochaitus*), were hunted to extirpation.

THREATS

The Knysna–Tsitsikamma Forest is threatened by the same forces that are at work all over the globe: increasing human population, human economics, and climate change. Golf courses, vacation villas, and eco-farms crowd South Africa's southern coast, requiring ever-increasing road and infrastructure development. The primary highway along the coast has been expanded and rerouted several times, further fragmenting indigenous forest patches. Only roughly 4 percent of the region is strictly protected (see table 10-1 in chapter 10).

Tree Harvest

One of the biggest problems facing Knysna–Tsitsikamma Forest conservation is that conservation planning and active wood-resource extraction are taking place in a knowledge vacuum. There is little information about wildlife population levels, habitat use, or its influence on forest dynamics. Few studies have explored whether forest patches are connected, and at what spatial scales, or the relationship between forest and adjacent fynbos.

The Knysna–Tsitsikamma Forest is the only indigenous South African forest harvested on a commercial scale. South Africa's National Forests Act (Act No. 84 of 1998) strives to "maintain and enhance the long-term health of forest ecosystem(s), while providing ecological, economic, social and cultural opportunities for the benefit of present and future generations." To achieve these results, South African foresters, for several decades now, have selected dying, senescent, or windthrown trees for harvest. The idea is to imitate the formation of small gaps from individual tree death. However laudable this logging method, it still raises some concerns. Besides creating a network of skid roads with their own inherent problems, removing large-diameter senescent trees reduces habitat for cavity-nesters and snag-dependant wildlife. In addition, many key forest trees are slow-growing; it may take centuries to replace lost forest biomass (Midgley and Seydack 2006).

Plantation Fire Management

Fast-growing, nonnative pine (*Pinus* spp.) and Eucalyptus (*Eucalyptus* spp.) plantations supply most of South Africa's wood and pulp; plantations are widespread across the Knysna–Tsitsikamma region. Most plantations are planted in fynbos and grassland immediately adjacent to indigenous forest patches, but Hoffman (1997) estimated that just over 21 percent of the Knysna–Tsitsikamma Forest has been converted to plantations. Similarly, Berliner and Desmet (2007) estimated that 25 percent of the forest has been converted or degraded. Besides the obvious loss of forest habitat, introduction of invasive species, and surface-water loss (Berliner and Desmet 2007), plantation management near the mountains requires setting intentional fires to protect valuable plantations from wildfire. Prescribed fires burn indigenous forest patches all too frequently, causing major shifts in understory composition (Geldenhuys 1994). Indeed, the Cape Action Plan for the Environment, a regional conservation planning initiative, identifies "poor fire management" as one of the six key issues causing biodiversity loss (Younge 2002).

Climate Change

Recent climate change models predict that the Knysna–Tsitsikamma region faces significantly warmer and drier conditions (Midgley et al. 2003). As much as a fifth of the protected forest in the former Tsitsikamma National Park could disappear in just 50 years (Hulme 1996). Climate change also threatens the persistence of fynbos that surrounds many forest patches (Midgley et al. 2003). On the other hand, increasing carbon dioxide levels may compensate for greater water stress and thus benefit the forests to a degree (Midgley and Seydack 2006).

CONSERVATION PRIORITIES

The Cape Action Plan for the Environment (CAPE—Younge 2002) and the Eastern Cape Biodiversity Conservation Plan (ECBCP—Berliner and Desmet 2007) both aim to protect regional biodiversity, incorporate ecological pathways and processes, and integrate economic opportunities to address otherwise overwhelming social issues. Both CAPE and ECBCP rely on coarse planning filters to identify areas for conservation, which risks missing the smaller portions of a spatially heterogeneous plant community like the Knysna-Tsitsikamma Forest. In the face of impending climate change, conservation of forest patches depends on both the preservation of species and the maintenance of ecosystem processes within and among forest patches (Midgley et al. 1997), as well as the surrounding fynbos matrix.

In March of 2009, the Garden Route National Park was created out of the Tsitsikamma National Park, the Wilderness National Park, and state lands. The new park includes estuaries, lowland fynbos, a long stretch of coastline, and a complex of indigenous forest patches. Within the forest patches, the management that was taking place prior to the formation of the park will continue. Former National Park forest will be completely protected (i.e., no tree removal), and former state land will continue to be selectively harvested as described above.

However, very little research has been done on ecological processes and pathways within the Knysna-Tsitsikamma Forest, or between the forest and adjacent plant communities (Midgely et al. 2003). Ecological processes are important because forest patches are not edaphically controlled, so patch locations could shift with anticipated climate change. On a more local scale, snag-dependent wildlife may play a critical role in perpetuating forest communities. As South Africa's population increases, timber and pulp needs will undoubtedly increase, putting even more pressure on the fynbos-forest ecosystem.

We recommend that the South African government or other sources provide funding to study ecological relationships within the Knysna-Tsitsikamma Forest and between the forest and adjacent biomes. This forest deserves as much conservation focus as does the famous fynbos growing along its edges. We believe that in so doing, South Africa will be well rewarded with a small but invaluable jewel in its conservation crown.

LITERATURE CITED

Berliner, D., and P. Desmet. 2007. *Eastern Cape biodiversity conservation plan:Technical report, no. 2005-012.* Department of Water Affairs and Forestry Project, Pretoria, South Africa.

Cowling, R. M. 1992. *Ecology of fynbos: Nutrients, fire and diversity.* Cape Town, South Africa: Oxford Univ. Press.

———, D. M. Richardson, and S. M. Pierce. 1997. *Vegetation of Southern Africa.* Cape Town, South Africa: Cambridge Univ. Press.

Geldenhuys, C. J. 1993. Floristic composition of the Southern Cape forests with an annotated check-list. *South African Journal of Botany* 59 (1):26–44.

———, C. J. 1994. Bergwind fires and the location pattern of forest patches in the Southern Cape landscape, South Africa. *Journal of Biogeography* 21 (1):49–62.

Goldblatt, P., and J. Manning. 2002. Plant diversity of the Cape region of southern Africa. *Annals of the Missouri Botanical Garden* 89 (2):281–302.

Graham, I. H. K., M. Landman, L. Kruger, N. Owen-Smith, D. Balfour, W. F. de Boer, A. Gaylard, K. Lindsay, and R. Slotow. 2008. Effects of elephants on ecosystems and biodiversity. Pp. 101–47 in *Elephant management: A scientific assessment of South Africa.* Ed. Robert J. Scholes and Kathleen G. Mennell. Johannesburg, South Africa: Witwatersrand Univ. Press.

Hoffman, M. T. 1997. Human impacts on vegetation. Pp. 507–34 in *Vegetation of Southern Africa.* Ed. R. M. Cowling, D. M. Richardson, and S. M. Pierce. Cape Town, South Africa: Cambridge Univ. Press.

Hulme, M., ed. 1996. *Climate change and Southern Africa: An exploration of some potential impacts and implications in the SADC region.* Gland, Switzerland: Climatic Research Unit, Univ. of East Anglia, Norwich, Great Britain; and World Wildlife Fund International.

Midgley, G. F., L. Hannah, D. Millar, W. Thuiller, and A. Booth. 2003. Developing regional and species-level assessments of climate change impacts on biodiversity in the Cape Floristic Region. *Biological Conservation* 112:87–97.

Midgley, J. J., R. M. Cowling, A. H. W. Seydack and G. F. van Wyk. 1997. Forest. Pp. 278–99 in *Vegetation of Southern Africa.* Ed. R. M. Cowling, D. M. Richardson, and S. M. Pierce. Cape Town, South Africa: Cambridge Univ. Press.

——— and A. H. W. Seydack. 2006. No adverse signs of the effect of environmental change on tree biomass in the Knysna forest during the 1990s. *South African Journal of Science* 102:96–97.

Schulze, R. E. 1997. Climate. Pp. 21–42 in *Vegetation of Southern Africa.* Ed. R. M. Cowling, D. M. Richardson, and S. M. Pierce. Cape Town, South Africa: Cambridge Univ. Press.

Scott, L., H. M. Anderson, and J. M. Anderson. 1997. Vegetation History. Pp. 62–84 in *Vegetation of Southern Africa.* Ed. R. M. Cowling, D. M. Richardson, and S. M. Pierce. Cape Town, South Africa: Cambridge Univ. Press.

van Daalen, J. C. 1981. The dynamics of the indigenous forest–fynbos ecotone in the Southern Cape. *South African Journal of Forestry* 119:14–23.

von Maltitz, G., L. Mucina, C. Geldenhuys, M. Lawes, H. Eeley, H. Adie, D. Vink, G. Fleming, and C. Bailey. 2002. *Classification system for South African indigenous forests: An objec-*

tive classification for the Department of Water Affairs and Forestry. Pretoria, South Africa: Environtek.

Younge, A. 2002. An ecoregional approach to biodiversity in the Cape Floral Kingdom, South Africa. Pp. 168–88 in *Biodiversity, sustainability, and human communities: Protecting beyond the protected.* Ed. T. O'Riordan and S. Stoll-Kleemann. Cambridge: Cambridge Univ: Press.

CHAPTER 10

Crosscutting Issues and Conservation Strategies

Dominick A. DellaSala, Paul Alaback, Lance Craighead,
Trevor Goward, Holien Håkon, James Kirkpatrick, Pavel V. Krestov,
Faisal Moola, Yukito Nakamura, Richard S. Nauman, Reed F. Noss,
Paul Paquet, Katrin Ronneberg, Toby Spribille, David Tecklin,
and Henrik von Wehrden

Throughout this book we have made the case that temperate and boreal rainforests are unique ecosystems globally and therefore worthy of stepped-up efforts for their study and conservation. Because entire ecosystems (and not just species) are at the brink of extinction (e.g., European relicts, old-growth coastal redwoods), a comprehensive vision is needed to ensure the persistence of rainforests in a time of unprecedented change brought about by the cumulative effects of climate change and ongoing land use.

We now turn our attention to the status and condition of temperate and boreal rainforests as seen through the lens of the crosscutting conservation issues discussed initially in the regional chapters and now synthesized into actionable steps for conservation groups, decision makers, and the public. We provide basic conservation approaches to support and help guide coordinated conservation actions around the world. (See www.temperaterainforests.org for coordinated temperate rainforest actions already underway.) We offer a comprehensive set of principles to help guide conservation groups and decision makers concerned about the fate of these rainforests.

A 360-DEGREE VIEW OF TEMPERATE AND BOREAL RAINFORESTS

When it comes to conservation planning, three things matter above all: evolution, land-use history, and geography. Here, we summarize overall patterns in how each of these factors influences distribution patterns of rainforest communities across global hemispheres and geographic regions.

North vs. South

Unlike tropical rainforests, temperate and boreal rainforests in the Northern and Southern Hemispheres have evolved in isolation from one another. This has resulted in significant differences in their respective histories and ecological sensitivity.

One key characteristic of temperate rainforests of the Southern Hemisphere is a high degree of local endemism, that is, of species restricted to only one or a few regions (e.g., Valdivia, chapter 5; Australasia, chapter 8). This fact alone makes many species in these ecosystems extremely vulnerable to habitat loss and to any factor that effectively compromises existing protected areas. The loss of species through displacement and habitat fragmentation is well documented here, dating back even to glacial periods during the Pleistocene (~1.8 million to 10,000 years ago), when fragmentation resulted in local or even regional extinction (e.g., Premoli et al. 2000). In light of this, glacial relicts in particular should be high priorities for conservation. Current land-use practices often exacerbate these alarming trends, as, for example, in the case of southern Chile, where severe fragmentation from logging has effectively isolated coastal forests from other forest remnants (Armesto et al. 1998; Smith-Ramirez et al. 2004, 2005).

Many southern rainforests have evolved with a very low frequency of fire (e.g., Australasia, Valdivia). Hence, the introduction of fire and fire-adapted exotic species such as pines (*Pinus* spp.) can have ecological consequences much more profound and long-lasting than has occurred in the Northern Hemisphere (Silla et al. 2003; Veblen et al. 2008). The introduction of ponderosa pine (*P. ponderosa*) in Argentina, for example, now promotes wildfires much hotter and more ecologically destructive than would have occurred naturally (see Kitzberger et al. 1997). In Chile and Argentina, the intentional introduction of fire as a "management tool" has done tremendous damage to the iconic *alerce* (*Fitzroya cupressoides*) forests (see Veblen et al. 2008).

Lastly, long historic isolation often results in an inability to withstand invasion from exotic species. Not surprisingly, many rainforests of the Southern Hemisphere are exceptionally vulnerable to plant and animal introductions from temperate regions of the Northern Hemisphere (Arroyo et al. 2000). Because of the history of isolation and many related factors, Southern Hemisphere rainforests are generally considered more sensitive to climate change than their northern counterparts (e.g., Alaback and McClellan 1993; Mooney et al. 1993). Complex geology and lack of good dispersal from potential refugia in Patagonia are additional constraints on the adaptability of rainforest species to climate change.

Europe vs. the Rest

Historical differences in human settlement patterns in the Europe versus those of the rest of the world have had a profound effect on Europe's temperate and boreal rainforests. The long history of human occupation there has all but replaced rainforests with widely scattered, semi-natural relicts (see chapter 6), very few of which are strictly protected, and nearly all of which are affected by domestic livestock. European relicts are therefore on "life support" as most of the original species, structure, and processes were altered centuries ago by extensive human use. Those undertaking rescue efforts today operate much like detectives in search of clues, estimating earlier conditions through a careful examination of potential vegetation and paleobotany. The remaining relicts are indeed the building blocks for an uncertain but essential reconstruction process. The take-home message here is not to let the rest of the world's rainforests get to the point where Europe finds itself today.

The situation outside Europe is not nearly as dire yet, but unfortunately is rapidly headed that way. Remaining large, intact areas are mostly situated in northern regions (e.g., portions of British Columbia and Alaska, chapter 2; Interior British Columbia northward, chapter 3; Valdivia, chapter 5). Across the globe, conservation has focused mainly on intact forests (e.g., roadless areas in Alaska, British Columbia, Chile), large carnivores and Pacific salmon (e.g., the Great Bear Rainforest in British Columbia), focal species (several in Chile), and threatened species (the northern spotted owl [*Strix occidentalis caurina*], marbled murrelet [*Brachyramphus marmoratus*] in the Pacific Northwest; mountain caribou [*Rangifer tarandus*] in Northwestern North America). Additional efforts include restoring degraded ecosystems (clear-cuts or plantations, riparian areas, watersheds in the Pacific Northwest and Europe).

WHAT REMAINS AND HOW MUCH SHOULD WE PROTECT?

These fundamental questions are, of course, very much at the heart of all conservation strategies globally, and they apply equally to temperate and boreal
rainforests. In the regional chapters we provided protected areas estimates based
on different mapping methodologies and definitions of protected areas (which
are still not standardized). Here, we turn our attention to the world database on
protected areas (UNEP-WCMC 2007), the only standardized global data set
with which to compare rainforest regions in terms of amount and type of protection. In addition, what counts as "protected" has been discussed throughout
this book (see chapter 3), and we have made the case for preferring that protected-areas inventories be based on IUCN Categories I and II, as these areas
are managed strictly for nature conservation, wilderness, scientific value, and
ecological restoration (examples include national parks and wilderness, Dudley
2008). Additional lower levels of protection do count, however, in local conservation, but unless otherwise noted these are not assessed here because of the
numerous exemptions that permit varying levels of development.

Twelve Percent as a Floor, Not a Ceiling

In 1987, the United Nations Commission on Environment and Development
commissioned a report that generally recommended tripling the global expanse
of protected areas (Brundtland Commission 1987). Most governments have in-
terpreted this as roughly 12 percent of every ecoregion of the world to receive
some level of protection. Since roughly 13 percent of the definitive rainforest
regions and about 11 percent of rainforest outliers are strictly protected (see
table 10-1), it would appear on face value that protection of these rainforests
does meet the global targets. However, putting additional areas in strict protection and taking other conservation measures will be necessary to meet representation targets as outlined by regionally specific conservation-area designs
(e.g., see discussions in chapters 2 and 3). Without considering representation
targets, one can incorrectly conclude that adequate protection levels have been
secured for rainforests, particularly if protected-areas targets are based on arbi-
trarily defined percentages (e.g., 12 percent). This is especially true if protected
areas disproportionately represent areas of low economic value (e.g., high elevation—so-called rock and ice), while excluding productive low-elevation
forests, which is the case for many protected-areas designations (Scott et al.
2000, DellaSala et al. 2001).

Using the reserve design concepts discussed throughout this book as well as
the regional targets provided by representation analyses, we conclude that about

Table 10-1. Extent, amount, and type of protection, as well as conservation targets for temperate and boreal rainforests.[a]

Region	Area (ha)[b]	IUCN I & II forest protections (ha)[c]	Conservation Targets
Pacific Coast of North America (conservation priority areas only)			
Chugach	1,485,349	1,260,000 (84.8%) of forested area	Manage tourism and off-road-vehicle use; solidify roadless-area protections.
Tongass	3,342,499	1,540,000 (46.1%) of forested area	Protect remaining old-forests and roadless areas—put at least 60% of region in strict protection.
Great Bear and Haida Gwaii	4,685,402 (adjusted to 7,400,000, see chapter 2)[c]	645,320 (13.8%); (adjusted to 2.6 million (35.1%), see chapter 2)[c]	Strictly protect 40–70% of rainforests to achieve representation and wildlife viability.
Clayoquot Sound	303,813	106,000 (34.9%)	Strictly protect 40–60%.
Pacific Northwest	4,775,981	235,750 (4.9%)	Strictly protect remaining old forests, roadless areas, and endangered species; restore salmon habitat.
Redwoods	877,396[d]	75,893 (8.7%)	Strictly protect remaining old-forest groves and focal species; focus on special elements, including landscape and watershed connectivity.
Pacific Coast Total[e]	27,274,225	3,862,963 (14.2%)	
Inland Northwestern North America (British Columbia only)	2,179,733	97,692 (4.5%)	Protect 45% of entire region—all remaining intact old forest.
Eastern Canada	5,969,641	319,972 (5.4%)	Identify and protect intact rainforests and focal species; control introduced moose populations; reduce air pollutants; limit new dam construction.

Table 10-1. Continued

Region	Area (ha)[b]	IUCN I & II forest protections (ha)[c]	Conservation Targets
Valdivia			
Argentina	348,371; adjusted to 2,211,888, based on regional GIS mapping	68,849 (19.8%); adjusted using regional data sets of strictly protected areas to 614,242 (27.8%)[c]	Strengthen and finance national network of protected areas; protect focal species; impose limits on dams and logging of primary forest.
Chile	12,211,573; adjusted to 9,752,451, based on regional GIS mapping	2,162,218 (17.7%); adjusted using regional data sets to 1,326,800 (13.6%)[c]	
Europe			
Norway	4,887,739	52,214 (1.1%)	Protect seminatural wood-
Ireland and Republic of Ireland	1,578,545	25,781 (1.6%)	lands, restore others with native species; restrict
Great Britain	5,064,759	None	livestock; reintroduce ex-
Bohemian Forest	220,199	15,607 (7.1%)	tirpated species (e.g.,
Northeast Alps and Swiss Prealps	745,915	323 (0.04%)	beaver, large carnivores); connect and expand pro-
Southeastern Alps and Northwest Balkans	577,425	2,528 (0.4%)	tected areas.
Japan and Korea	8,295,241	409,691 (4.9%); none in Korea	Protect remaining natural forests; restore native forests.
Australasia			
New Zealand	5,458,170	1,694,488 (31%)	Contain invasive species; reinstate indigenous-style burning of the lowland treeless vegetation.
Tasmania	3,132,684	284,075 (9.1%); this number is much higher (68% non–*Eucalyptus*, 32% *Eucalyptus*—chapter 6) but not available in a GIS database	Protect mixed forests.

Table 10-1. Continued

Region	Area (ha)[b]	IUCN I & II forest protections (ha)[c]	Conservation Targets
Australia	55,989	All non-*Eucalpytus* forests protected (no figures available in the world protected-areas database)	Protect mixed forests.
Total for Rainforest Regions (rainforest distribution model)	78,000,209	8,996,401 (11.5%)	Greatly expand reserve network to achieve representation targets (e.g., 40–70% of intact areas).
Regional Adjustment for Great Bear and Valdivia rainforests[c]	80,119,202	10,661,056 (13.3%)	
Outliers[f]			
Colchic	3,000,000	237,135 (7.9%)	Protect ~one-quarter of remaining forests in both regions.
Hyrcanic	1,960,000	54,839 (2.8%)	
Russian Far East	6,800,000	652,588 (9.6%)	Establish new nature reserves and increase connectivity among reserves and within regions; protect remaining tiger populations.
Inland Southern Siberia	7,000,000	1,201,129 (17.2%)	
Knysna-Tsitsikamma	235,483	9,456 (4.0%)	Fund new research; expand protected areas.
Total Outliers	18,995,483	2,155,147 (11.4%)	Greatly expand protected areas to achieve representation.

[a]Based on the rainforest distribution model (MaxEnt.), World Database on Protected Areas (UNEP-WCMC 2007), and regional conservation strategies.

[b]Based on rainforest distribution model (MaxEnt) except where regional adjustments were necessary to update global protected areas data sets.

[c]The UNEP-WCMC (2007) global protected-areas data set was the only standardized data set available for protected areas comparisons among regions and globally. Regional GIS-derived estimates, however, were used to supplement the database for both forest cover and IUCN I& II (strictly protected) areas in order to correct for more recent protected-areas additions not yet available in the global data set. In such cases, regional forest-cover totals were used instead of Maxent totals for standardization purposes.

Table 10-1. Continued

ᵈObtained from the California Department of Forestry: www.frap.cdf.ca.gov/data/frapgisdata/ download.asp?rec=redwood/. Redwood totals were not included in the Pacific Coastal summation due to overlap with the Pacific Northwest region.
ᵉAdjusted to include additional areas outside conservation priorities that were not included in this table but were included in table 1-2 as part of the overall regional total.
ᶠ Not directly comparable to the rainforest distribution model due to differences in mapping methodologies.

50 percent of temperate and boreal rainforests globally will require some form of long-term protection, preferably strict. Conservationists can adjust this estimate regionally based on levels of intactness, sensitivities to human intrusion, and how much rainforest remains, particularly primary rainforest, through conservation-area design. Further, because the protected areas database for rainforests is incomplete at this time (UNEP-WCMC 2007), we recommend updating it to capture more-recent conservation agreements (e.g., as in British Columbia). Nevertheless, our estimates of strict protection are likely to be conservative (± a few percentage points) but still far lower than levels most conservation scientists recommend as necessary to sustain rainforest ecosystems. Notably, ongoing demand for wood products globally, especially in North America, Europe, and Japan, is increasingly putting unprotected rainforests at risk. Conservation is a global responsibility, and regional conservation the means to that end. Here, we scan the global horizon, region by region, to determine just how much of these forests is in strict protection and what needs to be done to unify conservation efforts globally.

The Pacific Coast of North America

Protection levels are higher in the more-remote northern region where there are more intact, old-growth areas, but overall protection totals about 14 percent of the region (see table 10-1). In the north, most of the Chugach National Forest (85 percent of the forest base) is strictly protected and this region, along with Australasia, is a global leader in rainforest protections. Inventoried roadless areas (over 2,000 hectares) on the Chugach are currently "protected" under the U.S. Roadless Conservation Rule, although the policy is under legal and administrative review (see Turner 2009 for legal and historical review). The Tongass (~46 percent of forest base), Great Bear and Haidi Gwaii (35 percent of forest base), and Clayoquot Sound (35 percent of forest base) are below but approaching representation targets (40–70 percent). In contrast, the Pacific Northwest USA had much lower levels (~5 percent) of strict protection (see table 10-1). Protecting remaining old forests and roadless watersheds in each of these regions would go a long way toward completing representation targets, particularly in places with high levels of intactness (e.g., Siskiyou Wild Rivers in southern Oregon, Great Bear and Tongass rainforests; see Strittholt and DellaSala 2001).

Inland Northwestern North America

This region still contains significant levels of older forests and intact areas (see chapter 3). However, in British Columbia, under 5 percent is strictly protected (all forest types; see table 10-1). Protection levels are generally well below regional conservation targets (45 percent). Government conservation strategies have focused mainly on mountain caribou—clearly an unsatisfactory emphasis given that many caribou populations are now facing extinction, and that caribou winter areas potentially remain open to logging and, in any event, to other development. Thus, protecting old-forest habitat and achieving a representative conservation network is vital for caribou survival as well as a broader suite of old-forest species (especially lichens), as recommended in regional conservation–area design.

South America (Valdivia Temperate Rainforests)

Despite having moderate (18–20 percent, see table 10-1) protection, forest cover in Valdivia has been reduced by 60 percent (see chapter 5), and few strictly protected reserves exist in the most biodiverse coastal rainforest types. Development has pushed many aquatic and forest ecosystems to the brink, primarily pressure from land conversion, logging, and hydroelectric dams. The network of protected areas needs strengthening, particularly through private financing, as there are few public lands.

Eurasia

Throughout Eurasia, protection levels, intactness, and remaining rainforest are very different, not surprisingly, from those in the Northern Hemisphere (British Columbia and Alaska). Protection of rainforests is less than 2 percent for most regions, with Great Britain having no strictly protected rainforests (see table 10-1). Best-off in terms of rainforest extent are central and southern Europe, though even few protected areas exist in these areas, with the exception of the Bohemian Forest (7 percent). Japan, while densely forested, has protected just 5 percent of its semi-natural forests. There are no strictly protected areas on the Korean Peninsula.

Australasia

Except for mixed forests of Tasmania and Australia where logging of *Eucalyptus* still occurs, rainforest protection is among the highest globally (see table 10-1). Tasmania and the South Island of New Zealand still have relatively large intact areas. Conservation strategies include protecting intact areas and including mixed forests of *Eucalyptus*, which we recommend be treated as rainforest with the same protection afforded all the region's primary forests (see chapter 8). At risk are the world's tallest hardwood trees, most of which are turned into wood chips for export.

Rainforest Outliers

We considered outliers separately because differences in mapping methodologies and databases rendered comparisons among regions unreliable. Nonetheless, we did apply the world-protected-areas database to outliers so that at least the protected-areas comparisons are drawn from a standardized data set.

Most (90 percent) of the lowland and foothill forests of the Colchic and significant portions of the Hyrcanic forests (40 percent) in the Western Eurasian Caucasus have been logged (see chapter 9), with few (less than 8 percent) areas strictly protected (see table 10-1). Regional conservation strategies will need to emphasize increasing protection levels to one-quarter of remaining forests. The situation in Inland Southern Siberia is not quite as dire, as most of the historic forest is still intact and about 17 percent protected. Logging in the Russian Far East, however, continues to chip away at the rainforest land base, with 60 percent of the region's forests gone (see chapter 9) and lower levels (less than 10 percent) protected. The Knysna-Tsitsikamma rainforests of South Africa are even worse off, as they are small and isolated, have few protected areas, and lack intactness (see table 10-1). Here, an entire ecosystem is at risk, and some species (e.g., elephants *Elephas maximus*) are already functionally extinct. Rainforest conservation strategies for each of these regions should involve new protected areas and strengthened enforcement of existing ones. At a minimum, for instance, this is needed to ensure viability of endangered snow leopard (*Panthera pardus orientalis*) and Ussuri tiger (*P. tigris altaica*) of the Sikhote-Alin forests, as these charismatic species are already "circling the drain."

GLOBAL RAINFOREST CONSERVATION VISION

Given the low level of protection and the dire state of most of the world's temperate and boreal rainforests, we recommend the adoption of a coordinated vision by conservation groups, based on fundamental conservation principles and approaches to climate change planning.

Conservation Principles

For decades, conservationists have worked to conserve ecosystems and threatened species guided by reserve-design concepts (Noss and Cooperrider 1994; Lindenmayer and Franklin 2002; Groves 2003). We repeat some fundamental concepts in the context of rainforest conservation, as they are key building blocks for a global vision:

1. *Conserve representative species and ecosystems in a cohesive landscape-level system of strictly protected reserves integrated with lands of varying land-use intensities* (see Noss and Cooperrider 1994; Lindenmayer and Franklin 2002). In many parts of the world, reserves often border villages or small towns where people are economically dependent on local natural resources. It follows that economic incentives and the integration of conservation practices into local communities are essential in order for reserves to function as more than "paper parks" and to be resilient to global environmental change (Brooks et al. 2006; Janssen and Ostrom 2006).

2. *Provide landscape connectivity to facilitate dispersal of wildlife among intact areas.* This is important in naturally isolated rainforests such as archipelagos (e.g., coastal North America, see chapter 2) where species may be especially vulnerable to fragmentation caused by land development. Barriers to species dispersal across landscapes in the form of roads, clearcuts, forest plantations, developed landscapes, and hydroelectric dams, among other uses, can lead to fragmentation, genetic isolation, and population declines of sensitive species (Lindenmayer and Fischer 2006). While documentation of such effects is usually restricted to charismatic wildlife species, there is growing evidence that fragmentation and isolation of wildlife populations and habitat is equally important to other rainforest biota, including lichens, mosses, many invertebrate species, and amphibians (Fenton and Frego 2005; Ewers and Didham 2006; Lindenmayer and Fischer 2006).

3. *Maintain viable populations of native wildlife.* Throughout this book we have discussed focal species conservation as a cornerstone of conservation approaches (e.g., see chapters 2, 3, 5). We emphasized the maintenance of viable populations of keystone or indicator species, including, for example, sensitive lichens, caribou, salmon, and northern spotted owl.

4. *Allow evolutionary and ecological processes to continue to shape rainforest communities.* While island biogeography plays a pivotal role in shaping many coastal rainforests, human-imposed fragmentation of wildlife habitat often exceeds the adaptive capabilities of rainforest species. Intact areas, especially on archipelagos, provide a necessary buffer that allows sensitive wildlife to adapt to the cumulative impacts of land use. Additionally, places where endemic species (or subspecies) are especially concentrated (such as Chile and Argentina) are important in adaptive radiations, and many endemic species are vulnerable to extirpations, given narrow distributions and often-limited dispersal capabilities.

Conservation Recommendations

With the exception of non-*Eucalyptus* rainforests in Australia and Tasmania, temperate rainforests in New Zealand, and Alaska's Chugach National Forest, protection levels are far too low to ensure that rainforest communities will continue to provide the critical ecosystem services upon which all life depends. While some may feel that there are few if any rainforests in Europe, remaining relicts, particularly in central Europe, are valuable as restoration blueprints (see chapter 6). In addition, we call attention to important discoveries in regions not widely regarded as rainforests but which should be part of rainforest conservation (e.g., the Caucasus, the Balkans, the Alps, Eastern Canada, the Russian Far East, Southern Siberia, and South Africa).

- *Expand the number and size of protected areas.* In places where significant levels of intactness remain (e.g., North America), protection (strict) levels should be set at 40–70 percent of the region, as recommended by most scientists.
- *Protect all remaining mature and old-growth rainforests.* There are many reasons for protecting these forests, most notably as habitat for wildlife that require such ecosystems, but also because of their ability to store carbon for centuries (see chapter 11). Classic examples include the Tongass, Great Bear, Inland Northwest of North America, Chile, and Australasia.
- *Maintain intact areas in roadless condition.* Areas without roads are increasingly rare (for North America see Heilman et al. 2002; Ritters and Wickham 2003; DellaSala 2007), and nearly nonexistent throughout most of Europe. Roadless areas, on the other hand, are repositories of biological diversity (e.g., DeVelice and Martin 2001; Strittholt and DellaSala 2001; Loucks et al. 2003), contain relatively intact ecological processes (DellaSala and Frost 2001), and generally have low levels of exotic-species

invasions (Gelbard and Harrison 2005). There is also evidence that eco-system services provided by intact areas (e.g., recreation) have important economic value for human communities (Southwick Associates 2000; also see Niemi et al. 1999 for economic value of ecosystem services of old-growth forests in the Pacific Northwest).

- *Protect remaining semi-natural woodlands in places where rainforests are especially scarce.* Rainforest relicts, isolated though they may be, are essential to restorative actions throughout Europe. These areas should be protected from domestic livestock to allow native plant communities sufficient time to recover.

- *Restore threatened species, ecosystems, and ecosystem processes.* In some regions (e.g., Europe), restoration of semi-natural conditions affords the best shot at re-creating some semblance of rainforests. For others (e.g., British Columbia, Tasmania, Chile, coastal Alaska), it is most prudent to protect sufficiently large areas in order to avoid the expense and risk of restoring degraded forests later. Restoration, however, may involve more than returning species to their former distributions, as in some cases (e.g., Australia) it needs to be achieved by changing management (e.g., fire regimes, logging practices) in areas adjacent to rainforests in order to avoid insularity effects.

- *Develop or implement regionally specific conservation.* Conservation-area design, focal-species conservation, representation, and GAP-analysis techniques (see chapters 2 and 3), among other methods, are important tools for conservationists and should form the scientific foundation of conservation strategies. Such strategies should be linked globally as part of a network of rainforest conservation actions (see, for example, www.temperate rainforests.org).

- *Provide incentives for landowners to participate in rainforest conservation.* In some regions (e.g., Chile), protected-areas conservation has been mainly left to land acquisitions, as there are few public lands. Other incentives might include compensation for forgoing timber harvest as part of climate change accords (see chapter 11).

- *Develop international accords on rainforest conservation in nations with overlapping ecosystem boundaries.* (See chapter 11 for further details.) Rainforest ecosystems do not stop at political boundaries, and therefore we recommend managing them seamlessly through coordinated actions. Examples include coastal and inland rainforests of North America, which span multiple jurisdictions (Canadian provincial and federal governments, U.S. state and federal governments, indigenous groups, private forest license holders), rainforests of Chile and Argentina, and European relicts. Greater conservation benefits often result when transborder parks or reserves are jointly established, and cooperative or incentives-based agreements are forged.

- *Increase monitoring, research, and public outreach on the importance of temperate and boreal rainforests.* Throughout this book, we argue that these rainforests have not received nearly as much attention as their tropical counterparts. Public outreach should raise awareness of the importance of temperate and boreal rainforest through greater networking among rainforest conservation groups. Moreover, research and monitoring linking temperate and boreal rainforests around the world are vital, particularly in areas not widely regarded as rainforest (e.g., the Caucasus, the Russian Far East, Knysna-Tsitsikamma, Eastern Canada).

- *Link markets campaigns regionally and globally to limit consumption of wood products coming from endangered regions or intact and older forests.* Market campaigns (see www.forestethics.org/markets-campaigns) use consumer pressure to deflect consumption of wood products away from threatened areas. When combined with techniques developed by scientists to identify high-conservation-value forests (e.g., Strittholt et al. 2007), such approaches are indeed effective in shifting consumption. For instance, conservation groups (e.g., Forest Ethics, www.forestethics.org) have successfully secured protection for 26 million hectares of vulnerable rainforests globally through public-education campaigns and forest-products boycotts. Campaigns are especially effective when joined with efforts to reduce wasteful consumption and shift remaining consumption to wood products produced in an ecologically and socially responsible manner (e.g., the Forest Stewardship Council; www.fsc.org). As part of this effort, we recommend an analysis of trade in timber products across rainforest regions to determine additional areas where outreach and education can provide the biggest bang for the conservation buck.

AN UNCERTAIN FUTURE

What lies ahead for temperate and boreal rainforests is murky at best. Clearly, rainforests will continue to exist in places strictly protected from land-use impacts. Even so, it seems fair to ask whether even the protected forests will continue to sustain the rainforest communities that define them today when their surroundings are increasingly developed. As humanity approaches 9 billion people at the turn of this century, what will this mean for critical ecosystem services like clean water, carbon storage, and abundant fish and wildlife populations we have come to expect from rainforests?

Even in times of pending crisis, there is opportunity. While ecosystems are unraveling globally (see Hassan et al. 2005) and climate disruptions will likely be either severe or catastrophic, depending on the level at which greenhouse gas emissions level off (see chapter 11), rainforest citizens and governments have a unique opportunity to change direction while there is still time. Although plans for triggering such needed action remain elusive, the conservation vision described herein ought to become part of international treaties and accords on temperate and boreal rainforests joined with similar efforts by conservation groups working to protect tropical rainforests and other endangered ecosystems.

Only time will tell whether the critical ecosystem services provided by rainforests will be appropriately valued for the sum-of-their-ecosystem-parts rather than extractive uses. Clean water, clean air, viable fish and wildlife populations, productive soils, long-term carbon storage, and other critical services will one day overshadow timber or other interests—as they already do, of course, whether or not most people and governments have yet become sensitive to their importance. Only then will these rainforests take their place on a global stage where they so rightfully belong. Their fate in years to come is really a question of how much political will and conservation commitment the world is willing to invest in global conservation in order to ensure an enduring legacy for generations to come.

LITERATURE CITED

Alaback, P. B., and M. McClellan. 1993. Effects of global warming on managed coastal ecosystems of western North America. Pp. 299–327 in *Earth system response to global change: Contrasts between North and South America*. Ed. H. A. Mooney, E. Fuentes, and B. I. Kronberg. New York and San Diego: Academic Press.

Armesto, J. J., R. Rozzi, C. Smith-Ramirez, and M. T. K. Arroyo. 1998. Conservation targets in South American temperate forests. *Science* 282:1271–72.

Arroyo, C. Marticorena, O. Matthei, and L. Cavieres. 2000. Plant invasions in Chile: Present patterns and future predictions. Pp. 385–421 in *Invasive species in a changing world*. Ed. H. A. Mooney and R. J. Hobbs. Washington DC: Island Press.

Brooks, J. S., M. A. Franzen, C. M. Holmes, M. N. Grote, and M. B. Mulder. 2006. Testing hypotheses for the success of different conservation strategies. *Conservation Biology* 20:1529–38.

Brundtland Commission. 1987. *Our common future: Report of the World Commission on Environment and Development*. Oxford: Oxford Univ. Press.

DellaSala, D. A. 2007. Where no road has gone before: Why roadless areas are nature's ark. Pp. 105–12 In *Thrillcraft.* Ed. G. Wuerthner. Washington, DC: Island Press.

———, and E. Frost. 2001. An ecologically based strategy for fire and fuels management in National Forest roadless areas. *Fire Management Today* 61 (2):12–23.

———, N. L. Staus, J. R. Strittholt, A. Hackman, and A. Iacobelli. 2001. An updated protected-areas database for the United States and Canada. *Natural Areas Journal* 21:124–35.

DeVelice, R. L., and J. R. Martin. 2001. Assessing the extent to which roadless areas complement the conservation of biological diversity. *Ecological Applications* 11 (4):1008–18.

Dudley, N., ed. 2008. *Guidelines for applying protected areas management categories.* Gland, Switzerland: IUCN.

Ewers, R. M., and R. K. Didham. 2006. Confounding factors in the detection of species responses to habitat fragmentation. *Biological Reviews* 81:117–42.

Fenton, N. J., and K. A. Frego. 2005. Bryophyte (moss and liverwort) conservation under remnant canopy in managed forests. *Biological Conservation* 122:417–30.

Gelbard, J. L., and S. Harrison. 2005. Invasibility of roadless grasslands: An experimental study of yellow star thistle. *Ecological Applications* 15 (5):1570–80.

Groves, C. R. 2003. *Drafting a conservation blueprint: A practitioner's guide to planning for biodiversity.* Washington, DC: Island Press.

Hassan, R., R. Scholes, and N. Ash, eds. 2005. Ecosystems and human well-being: Current state and trends. *Millennium Ecosystem Assessment,* Vol. 1. Washington DC: Island Press.

Heilman, G. E., Jr., J. R. Strittholt, N. C. Slosser, and D. A. DellaSala. 2002. Forest fragmentation of the conterminous United States: Assessing forest intactness through road density and spatial characteristics. *Bioscience* 52 (5):411–22.

Janssen, M. A., and E. Ostrom, eds. 2006. Resilience, vulnerability, and adaptation: A cross-cutting theme of the International Human Dimensions Programme on global environmental change. *Global Environmental Change* (Special Issue) 16:235–316.

Kitzberger, T., T. Veblen, and R. Villalba. 1997. Climatic influences on fires regimes along a rain forest-to-xeric woodland gradient in northern Patagonia, Argentina. *Journal of Biogeography* 24:35–47.

Lindenmayer, D. B. and J. F. Franklin. 2002. *Conserving forest biodiversity.* Washington, DC: Island Press.

———, and J. Fischer. 2006. *Habitat fragmentation and landscape change: An ecological and conservation synthesis.* Washington, DC: Island Press.

Loucks, C., N. Brown, A. Loucks, and K. Cesareo. 2003. USDA Forest Service roadless areas: Potential biodiversity conservation reserves. *Conservation Ecology* 7 (2):5. www.consecol.org/vol7/iss2/art5/

Mooney, H. A., E. Fuentes, and B. I. Kronberg, eds. 1993. *Earth system response to global change: Contrasts between North and South America.* New York and San Diego: Academic Press.

Niemi, E., E. Whitelaw, and A. Johnston. 1999. *The sky did not fall: The Pacific Northwest's response to logging reductions.* Eugene, OR: EcoNorthwest.

Noss, R. F., and A. Y. Cooperrider. 1994. *Saving nature's legacy.* Washington, DC: Island Press.

Premoli, A. C., T. Kitzberger, and T. T. Veblen. 2000. Isozyme variation and recent biogeographical history of the long-lived conifer *Fitzroya cupressoides. Journal of Biogeography* 27:251–60.

Ritters, K. H., and J. D. Wickham. 2003. How far to the nearest road? *Frontiers in Ecology and Environment* 1 (3):125–29.

Scott, M. J., F. W Davis, R. G. McGhie, R. G. Wright, C. Groves, and J. Estes. 2000. Nature reserves: Do they capture the full range of America's biological diversity? *Ecological Applications* 11 (4):999–1007.

Silla, F., S. Fraver, A. Lara, T. Allnutt, and A. Newton. 2003. Regeneration and stand dynamics of *Fitzroya cupressoides* (Cupressaceae) forests of southern Chile's central depression. *Forest Ecology and Management* 165:213–24.

Smith-Ramirez, C. 2004. The Chilean coastal range: A vanishing center of biodiversity and endemism in South American temperate rainforests. *Biodiversity and Conservation* 13:373–93.

———, J. J. Armesto, and C. Valdovinos, eds. 2005. *Biodiversidad y ecología de los bosques de la Cordillera de la Costa, Chile.* Santiago, Chile: Editorial Universitaria.

Southwick Associates. 2000. Historical economic performance of Oregon and Western counties associated with roadless and wilderness areas. www.onrc.org/info/econstudy/

Strittholt, J. R., and D. A. DellaSala. 2001. Importance of roadless areas in biodiversity conservation in forested ecosystems: A case study—Klamath-Siskiyou ecoregion, USA. *Conservation Biology* 15 (6):1742–54.

———, N. L. Staus, G. Heilman Jr., and J. Bergquist. 2007. Mapping high-conservation-value and endangered forests in the Alberta foothills using spatially explicit decision-support tools. Conservation Biology Institute, Corvallis, Oregon.

Turner, T. 2009. *Roadless rules: The struggle for the last wild forests.* Washington, DC: Island Press.

UNEP-WCMC. 2007. World data base on protected areas GIS. December 20, 2007. www.wdpa.org/

Veblen, T., T. Kitzberger, E. Raffaele, M. Mermoz, M. Conzalez, J. Sibold, and A. Holz. 2008. The historical range of variability of fires in the Andean-Patagonian Nothofagus forest region. *International Journal of Wildland Fire* 17:724–41.

CHAPTER 11

A Global Strategy for Rainforests in the Era of Climate Change

John Fitzgerald, Dominick A. DellaSala, Jeff McNeely, and Ed Grumbine

Like their tropical counterparts, temperate and boreal rainforests arose from a tightly knit association with climate (see chapter 1). Global climate disruptions are therefore likely to result in dire consequences to rainforests, particularly as temperature and precipitation levels are affected by climate change as discussed in the earlier chapters. Notably, declines in snow pack (thereby affecting water supply and aquatic organisms—see Mote et al. 2005) and recent increases in the duration of the fire seasons (Westerling et al. 2006) are already affecting regions with temperate and boreal rainforests. Particularly vulnerable are food-web dynamics involving woodland caribou and rainforest lichens (see chapter 3) and large carnivores and salmon (see chapter 2). The loss of keystone taxa at lower levels (lichens, salmon) in temperate and boreal rainforests may ultimately create the "perfect storm" whereby climatic thresholds for keystone species reverberate across food chains (e.g., see Lichatowich 1999 for how the loss of salmon could affect food-web dynamics in the Pacific Coast of North America).

In this closing chapter, we discuss the pivotal role that forests play as part of the planet's climate-control center, and recommend how they should be managed in a changing climate to prepare for and lessen climate-related impacts. We emphasize the conservation of biological diversity and critical ecosystem services, including the importance of long-term storage of carbon. In terms of tonnes per hectare of stored carbon, old-growth forests are carbon-dense ecosystems (Luyssaert et al. 2008; Hudiburg et al. 2009), and temperate rainforests are among the most carbon-dense ecosystems on the planet (Smithwick et al. 2002; Keith et al. 2009).

Many of our recommendations for forests are global in scope and most therefore require new, or changed, international agreements and implementation. The conservation community and interested scientists should organize around these broad-based conservation efforts that we consider central to the ongoing international discussions of a global climate-change accord, other arrangements or agreements, and rainforest protections generally.

Tropical, and to a lesser extent, temperate and boreal rainforests process (through photosynthesis and respiration) about twice the annual amount of the world's fossil-fuel emissions (Phillips et al. 2009). However, this critical ecosystem service is showing signs of reversal due to drought-related climate stresses slowing tree growth in the tropics (Turner et al. 2007; Phillips et al. 2009) and increasing tree mortality in the temperate zone (van Mantgem et al. 2009). Deforestation rates globally are second only to the burning of fossil fuels as an anthropogenic source of greenhouse gases, and this could trigger a dangerous feedback with climate change (IPCC 2007) whereby forests switch from carbon sink to carbon source (Phillips et al. 2009). Thus, a global response is urgently needed, one that includes both mitigation (through carbon storage and reductions in greenhouse-gas emissions) and climate-readiness (or "adaptation") in order to ensure that all the world's rainforests (tropical, temperate, and boreal) continue to provide the life-giving services we depend upon.

Although the evidence for global climate change is unequivocal and is largely human caused (IPCC 2007), policy responses have been slow. The longer real action at the necessary scale is delayed, the greater the procrastination penalty that will be paid by this and future generations. We begin by discussing why governments should adhere to broadly supported and enforceable global climate-change accords with major reductions in greenhouse-gas pollutants, particularly by reducing emissions from deforestation and degradation of the world's forests, and that they also adopt related policies and practices to secure lasting protections for forests and other ecosystems. Below we present eleven climate-change principles that are global in scope and therefore could form the foundation for additional accords, as well as domestic actions, to include the world's temperate and boreal rainforests in efforts to mitigate and prepare for climate change. The following principles are based largely on climate-change policy principles developed by the same authors, and adopted and delivered by the Society for Conservation Biology (2009) to the host ministers of the Copenhagen Conference of the Parties to the United Nations

Framework on Climate Change (UNFCCC). They also include excerpts from congressional testimony at climate-change hearings in advance of the climate-change summit (DellaSala 2009). In sum, the longer we wait to act on these essential remedial principles, the closer we will get to pushing a significant part of the world's climate control center over the edge.

1) *As soon as possible, create systems and policies that will reduce greenhouse gas concentrations to levels approaching historic levels, and no higher than 350 parts per million of carbon dioxide equivalent* (Hansen et al. 2008). Rapidly accelerating changes attributable in large part to climate change and its drivers include: losses of biological diversity along with changes in species' ranges and numbers; melting of glacial and polar ice; reductions in tree growth in the tropics (Clark et al. 2003; Feeley et al. 2007; Phillips et al. 2009); increased tree mortality (van Mantgem et al. 2009) and reduced tree growth (Turner et al. 2007) in the temperate zone; permafrost melting with concomitant release of methane; ocean acidification; desertification and drought; extreme weather patterns; increasing rates of flooding; and forest fires (Westerling et al. 2006; IPCC 2007). These changes are already being felt by the world's forests. Therefore, as of 2010, the level of greenhouse gases (about 392 parts per million of carbon dioxide and rising globally at a rate of 1.6 parts per million annually)[1]—or any level higher than that—is increasingly dangerous for society, for forests, and for the planet's life support systems. Policy responses should include:

- Proceed as quickly as possible to limit emissions of greenhouse gases and other driving agents, such as black soot, to safe levels (e.g., 350 parts per million of carbon dioxide).
- Act to restore degraded forest ecosystems to a healthy status, which is a primary defense against climate change.

2) *Cap and reduce emissions from every major sector, particularly forestry and agriculture, with rewards and consequences in proportion to performance* (Fitzgerald et al. 2009). Together forestry and agriculture account for over 30 percent of global greenhouse-gas pollutants (IPCC 2007; see figure 2.1—however, this does not include contributions from animal husbandry, i.e., livestock). When conserved and properly managed, natural ecosystems provide the most effective means for sequestration and long-term storage of carbon in growing plants. Thus, carbon-dense mature and old-growth forests in the Pacific Northwest and other primary forests should be conserved for this purpose at least. When these forests

---

[1]www.esrl.noaa.gov/gmd/ccgg/trends/index.html#global

BOX 11-1

Rainforests as Nature's Carbon Warehouse.

Given that the single largest threat to ecosystems worldwide is climate change, there has been a substantial amount of scientific work and debate on how forests affect the world's carbon budget, and, in turn, how the logging of forests releases greenhouse gases to the atmosphere. Plants in general and forests in particular play a large role in uptake and release of atmospheric carbon dioxide, and, in fact, they exchange considerably more of it each year than what humans put into the atmosphere through pollution (see Phillips et al. 2009). But the key point is that for plants there is a tight balance between carbon taken up through photosynthesis and carbon dioxide emitted through respiration. What is not necessarily so tightly balanced is the accumulation of carbon, especially in classical temperate rainforests where large trees, slowly decaying logs, and carbon-rich soils are the norm.

For years, timber-industry analysts have argued that cutting down old-growth rainforests was necessary for abating climate change. This is based on a specious claim that young trees (plantations) grow much faster than old ones in old-growth forests and the young trees are therefore pulling more carbon from the atmosphere. As long as you are looking at trees of equivalent sizes and not counting the rest of the forest, this is true. But several studies have shown that in fact, when you look at the entire forest and whole ecosystem, the opposite is true: logging of old-growth rainforest has led to more carbon dioxide released into the atmosphere, not less (Harmon et al. 1990; Luyssaert et al. 2008; Depro et al. 2008; Hudiburg et al. 2009). A net loss of stored carbon pools occurs in these forests through the release of carbon dioxide during site preparation following logging (e.g., burning and decomposition of slash), and the transport and manufacturing of wood products (Harmon et al. 1990; Harmon 2001; Law 2004). In addition, old trees continue to accumulate carbon for centuries, contrary to the view that they are carbon neutral (Luyssaert et al. 2008). Consequently, large amounts of carbon stored in old forests are not simply compensated for by planting fast-growing trees because forests are usually harvested again on short rotations (before they accumulate centuries of stored carbon in the original tree or forest), nor are they compensated for by storing some of the carbon in comparatively short-lived

BOX 11-1

Continued

wood products (Harmon et al. 1990; Harmon 2001; Law 2004). Notably, deforestation contributes globally to about 18 percent of greenhouse gas emissions, mostly in the tropics (IPCC 2007; Keith et al. 2009).

Temperate and boreal rainforests, since they generally are not subject to intense fires or other disturbances that emit large amounts of carbon dioxide, are among the most carbon-dense ecosystems on Earth (Keith et al. 2009). Especially dense concentrations of carbon have been documented in redwood forests of northern California, coastal rainforests of the Pacific Northwest, and mountain ash (*Eucalyptus regnans* the world's tallest hardwoods) forests of Australia (Keith et al. 2009). This is because unique genetics, evolutionary history, and suitable climatic conditions have resulted in extremely productive, fast-growing dense and tall forests in these regions. In other rainforest regions large carbon accumulations are expected as well, especially in boreal or sub-boreal rainforests where soils tend to accumulate the most carbon and in other forests with large trees that can grow at high densities, such as the alerce (*Fitzroya cupressoides*) in southern Chile. Despite the addition of many new temperate and boreal rainforest regions in this book and their high carbon density, however, the total amount of carbon contained in these rainforests is dwarfed by that of tropical rainforests because tropical forests occupy so much more land area. Nevertheless, when viewed together rainforests (temperate, boreal, and tropical) are truly among the world's carbon-storage champions, vital to the planet's climate-control center.

are cut down, a large portion (up to 40 percent by some estimates—Harmon et al. 1990, Harmon 2001) of their stored carbon is released as carbon dioxide (see box 11-1). Hence, a cap on forestry-related releases of carbon dioxide would go a long way toward reducing emissions from degradation and destruction of rainforests and should be considered by the UNFCCC, which might issue guidance on setting forestry emission caps as binding limits in any post-Kyoto protocol.

To reach the 350-parts-per-million-equivalent target for carbon dioxide emissions, each nation and each sector (e.g., power generation, commercial, industrial, agricultural, forestry, natural areas managers) should have annual greenhouse-gas-reduction and biological-sequestration and carbon-storage targets—with rewards and consequences in proportion to performance. One important step that can be taken in this regard is to adjust land-use practices, particularly forestry, to reduce emissions. Policy responses should include:

- Optimize carbon storage of forests through protection of carbon-dense old forests and long-rotation timber harvests.
- Adjust current agricultural subsidies to provide incentives for greenhouse-gas reductions, soil conservation, and stewardship practices.
- Require land-management agencies to use greenhouse gases as a metric for land-use decisions through environmental assessment and other laws and policies in order to cap emissions.

3) *Conserve and restore forests and other ecosystems, but recognize that forests will not be able to steadily offset increases in emissions* (Thompson et al. 2009). Although old-growth forests store carbon more effectively than most others systems, besides being deep repositories of biological diversity, they do have their limits, particularly when stressed by the combination of climate-change effects (e.g., drought, insect infestations) and land-use practices. Policy responses should include:

- Reduce both climate and non-climate stresses on forests and other eco-systems (e.g., logging, road building, livestock grazing), because fully functioning ecosystems play key roles in the overall carbon balance, and the species within them are likely to be more resistant and resilient to climate change (Noss 2001; Thompson et al. 2009).
- Provide significant domestic and international funding to restore degraded forest and other ecosystems in order to preserve their ability to sequester and store carbon for long periods, prevent releases of carbon dioxide from premature death or decay of trees, convert carbon dioxide to oxygen, and deliver the critical ecosystem services upon which all life depends.

4) *Phase out existing sources of greenhouse-gas emissions as quickly as possible, starting with the dirtiest first.* This will avoid and minimize negative effects to forests while maximizing the net positive impacts of improved environmental quality on ecosystem services and human health. For instance, in the energy

sector, such combinations include efficiency, demand management (e.g., re-forming utility rates to help lower-income users while encouraging higher-volume users to be more efficient), and the use of both renewable energy sources and the cleanest available fuels. Natural gas could be used as a transition fuel along with full assessments of alternatives and costs. Major economies and some developing nations already have several times the renewable-energy capacity that they need at practical prices when external costs and subsidies are considered. The chairman of the U.S. Federal Energy Regulatory Commission declared in 2009 that the United States is likely to need no new traditional base-load (coal or nuclear) power plants (see also, Barrett 2002; Hanson et al. 2004) if better efficiency standards and related initiatives are implemented. Policy responses should include:

- Replace highly polluting technologies with appropriate combinations of the best available technologies, determined transparently through environmental impact assessment and full life-cycle cost accounting.
- Adopt efficiency-first strategies that include revising energy and other natural resource prices, such as utility-rate structures, to reward efficiency.

5) *Be cautious about ratifying or enacting new measures that curtail or remove exist-ing domestic or international legal tools until the full implications are clarified.* Regard-less of any new climate agreements or laws, governments should apply existing conservation treaties, laws, and provisions (SCB 2008). First among these are the international conservation treaties to which most nations are parties or sig-natories. Also key are the development agreements and human rights principles that ensure that all the Earth's people are partners in efforts to create and sustain rainforests and communities. Domestic environmental laws should not be set aside in a rush to respond to climate change. In the United States, for example, the Clean Air Act and environmental assessment and wildlife laws can be ap-plied to climate change so as to benefit rainforests among other ecosystems. Several conservation treaties and agreements contain climate-relevant provi-sions that can be applied to rainforests and should be supported and empow-ered (McNeely 2009). For example, the community of nations has banned or strictly limited trade in products produced with unsustainable methods. Other agreements contain provisions that can also be applied: The General Agreement on Tariffs and Trade has recognized for over 50 years the right of individual na-tions to enforce higher conservation standards in their markets if the standards are applied fairly.

Policy responses should include:

- Maximize the use of existing legal and financial tools to reduce emissions by reordering priorities for government subsidies and spending, and by promulgating new regulations under existing laws such as the Clean Air and Water Acts, the National Environmental Policy Act, the Endangered Species Act and Federal land management laws, while new and improved tools are developed.
- Adopt new regulatory tools such as real-time public pollution monitoring and reporting, and taxes on any level of emissions, even those deemed legal, so that the communities affected will know immediately who is responsible for what pollution and will thus have some means of funding restoration and reparations.

6) *Fund forest and wildlife restoration and adaptation fully and directly.* Results of legislation passed and negotiation commitments made so far by the United States, for instance, indicate that some believe that domestic and international conservation objectives will be satisfied by devoting only a small percentage of the proceeds from the sale of pollution permits to forest restoration, setting up a few new international funds, and relying on carbon offsets. Such policies may be counterproductive if they create perverse incentives by selling permits to polluters in exchange for programs that help ecosystems adapt to even more pollution, which may not be possible if climate-change tipping points are reached. Developed nations, having benefitted from resources extracted from developing countries in a manner that has degraded forest and other ecosystems, have a responsibility both to provide additional funds for restoration and to stop themselves and others from practices that degrade biological diversity of forests, particularly primary and old-growth rainforest. Policy responses should include:

- Develop a major international effort to restore degraded rainforests and other ecosystems, to be funded in large part by developed nations.
- Encourage trade in forest products that is ecologically and socially responsible (e.g., Forest Stewardship Council certification).

7) *Use better science and enforcement to manage and limit the use of carbon offsets.* Those responsible for harm to forests sometimes choose, and are sometimes required, to make up for that harm. Many have done so by paying others to reduce emissions, or by paying for restoration projects to offset harm. However, offsets are complex and may be problematic. For instance, offsets could lead to "gaming the system," whereby landowners are paid to sequester carbon by

planting fast-growing seedlings that sequester carbon without accounting for losses to stored carbon pools from the logging of older forests on site. (Often this reduction in stored carbon is irreplaceable because repeat logging never allows forests to recoup the carbon originally stored in the uncut forest—see box 11-1.) The best approach is to avoid, minimize, and offset—and in that order, particularly if the offset is likely to be irregular, unreliable, or dependent on variables not controlled by the responsible party. Policy responses should include:

- Avoid relying on offsets as a pillar of government climate-change policy, unless the science in a specific case justifies such action and variables in governance and natural systems in that case can be adequately predicted and controlled.
- Encourage private and voluntary actors to use offsets responsibly by first and foremost reducing pollution and then offsetting what cannot be avoided.

8) *Practice stewardship in investment and procurement.* Several trillions of dollars in pension funds and other investments are now managed according to the United Nations' Principles for Responsible Investment. The UNEP Finance Initiative and many private services offer guidance in green investment, production, and procurement. Investing with forest stewardship in mind can deliver more than green technology. It can help all countries to meet international development goals in a sustaining and restorative manner—not only to have forest cover, clean air, and clean water, but to help ensure that children need not labor before learning and that women have the education, resources, and rights to determine the size of their families and the nature of their fates (O'Brien et al. 2009). Policy responses should include:

- Focus investments and spending so as to encourage measurable conservation progress and withhold investment from those who undercut this common cause.
- Enact and use new legal and financial tools, such as requiring companies and countries with publicly traded securities to disclose in their annual public reports to the Securities and Exchange Commission their environmental footprints, compliance with conservation treaties, potential liabilities and plans for improvements.
- Require the public employees and contractors' pension funds be screened for environmental impact and disallow matching investments

for pension purposes in companies or jurisdictions in the bottom half of the market in terms of environmental performance.

9) *Require the transparent use of natural and social sciences as well as law in setting and enforcing limits.* Investments in restoring human and environmental health that pay for themselves over the long term should be distinguished from apparent savings that vanish when full life-cycle costs are understood and accounted for (Nordhaus et al. 1999). A properly balanced climate program and budget will likely leave considerable sums for other social goods. Technologies used in the name of climate change that are riskier to ecosystems than other available production or efficiency technologies should be avoided. Treaties, statutes, and regulations under development should include mechanisms to provide legal standing to scientists and other citizens to enable them to pursue remedies for failed or inadequate compliance, such as those provided in existing U.S. law and the Aarhus and Nordic Conventions. Existing treaties, laws, and regulations should also be analyzed and, where appropriate, modified to ensure that such remedies exist for decision makers to use the best available science in decision making in a way that also applies the precautionary principle when there is significant scientific uncertainty. Independent scientists can also be encouraged to review and evaluate the efficacy of the measures taken. Policy responses include:

- Establish fundamental rules and systems that will respond to and incorporate the best available science and full-cost life-cycle accounting, particularly regarding forestry-related uptake and release of carbon dioxide.
- Include interim goals based on the precautionary principle and update them as quickly as possible in response to new scientific and technological information. In 1992, the global community, including the United States, expressed agreement on this principle through the Rio Declaration on Environment and Development. The precautionary principle is also well accepted in the scientific community as a means for preventing irreversible damage to natural resources, particularly when there is lack of scientific certainty.
- Apply the balanced use of legal limits and price signals, such as subsidies (or the elimination thereof), taxes on forestry-related emissions and/or sales of pollution allowances, well-informed resource management, regulations, and fair, comparable tariffs, with a portion of the proceeds—including tax and tariff rebates—that can be used for investments in renewable and restorative technologies.

10) *Lead a race to the top, not the bottom.* The transition to a carbon-neutral world is likely to be the biggest economic opportunity of the twenty-first century (Freidman 2008) and the most effective means for reducing climate-related stressors on forests. Numerous studies indicate that the countries that act first to build a low-carbon economy will reap economic benefits accordingly (Barrett et al. 2002; Freidman 2008). The urgency of climate change means that each party cannot wait to take action until most other parties act. The transition to the new low-carbon global economy can be a healthy one for forests and society. Climate change requires greater leadership from the countries with the greatest resources. A starting place can be to more aggressively apply existing treaties and elements thereof that have already been shown to work (e.g., the Montreal Protocol, the Convention on Biological Diversity, the Convention on Desertification, and the Convention on International Trade In Endangered Species—CITES) or that have not yet been enforced well enough to restore degraded forest ecosystems, reduce greenhouse gases, and create new jobs. Tariffs, incentives, taxes, aid, trade, and other tools should be designed to multiply their effectiveness domestically and internationally in order to support cleaner, safer energy production and forestry practices that also produce economic benefits. Research shows that we can multiply the impact of these tools by devoting significant portions of the proceeds of fees and the savings from efficiency to clean up production and otherwise address climate change (Barrett et al. 2002; Hanson et al. 2004). Policy responses include:

- Create improved systems that more effectively enforce legal standards based on the best-available science.
- Evaluate the effectiveness of measures taken and prescribe corrective or additional precautionary actions when warranted.

11) *Prepare forests and other ecosystems for climate change.* In addition to taking mitigation measures, we—all of us, and the ecosystems on which we depend—need to be prepared for the unavoidable consequences of climate change and ongoing pervasive ecosystem degradation. This can best be achieved by reducing existing environmental stressors such as logging, road building, and live-stock grazing, which impair the ability of forest species to adapt to climate change. Maintaining and restoring the properties of forests that allow them to be resilient or resistant to change is essential to their ability to adapt to human-induced stressors. Ecologists refer to *resiliency* as the ability of systems or species to rebound from perturbations, and *resistance* as being able to withstand change.

Forest ecosystems have many properties that allow them to rebound from and resist disturbance, such as the release of seed crops following a natural disturbance to the rainforest canopy and the presence of large, fire-resistant trees in fire-adapted forest types. Undeveloped floodplains and intact watersheds also are more resilient to floods, as they more gradually dissipate high flows, compared to developed floodplains, where flooding can have devastating ecological and economical consequences. Accelerating climate change and increasing land use are likely to overwhelm the capacity of rainforest species to rebound and resist disturbances, leading to undesirable ecosystem trajectories (Paine et al. 1998). This is particularly the case for tree plantations, which lack the diversity of species and genomes needed for forests to adapt to climate change. Policy responses include:

- Support management of both public and private lands to prepare ecosystems for climate change.
- Integrate existing conservation treaties, laws, and plans, such as state wildlife action plans (SWAPs) in the United States, directly into new climate-change agreements and laws so that the best science can guide and stimulate conservation action supported by new and additional resources.
- Protect intact roadless areas and old forests (genetic repositories), where ecosystems are most likely to have resilient and resistant properties, as climate refugia for wildlife dispersing in response to climate-forced migrations, and for critical ecosystem services, especially clean drinking water.

PARTING WORDS

The world is dangerously close to a tipping point on climate change (IPCC 2007). The steps we take in the coming years will dictate whether impacts are severe or catastrophic for both people and rainforest communities. This book provides a scientific foundation and a call to action for temperate and boreal rainforests that is only more urgent when viewed through the lens of climate change. Rainforests have persisted for millennia, adapting to subtle and sometimes rapid climate change but always within the productive capacity and evolutionary capabilities of rainforest species and communities. This time it is different, as rainforests face off against accelerated climate change combined with compounding stresses from humanity's ever-growing ecological footprint. There is still time to act but it will take bold actions from governments, organizations of all kinds, and citizens to move humanity closer to safe atmospheric levels of greenhouse-gas pollutants and more responsible forest-management and conservation practices. To do otherwise would trigger an unconscionable loss of countless species with which we have shared this remarkably alive planet for millennia.

LITERATURE CITED

Barrett, J. P., J. A. Hoerner, S. Bernow, and B. Dougherty. 2002. Clean energy and jobs: A comprehensive approach to climate change and energy policy. Economic Policy Institute, Washington, DC.

Clark, D. A., S. C. Piper, C. D. Keeling, and D. B. Clark. 2003. Tropical rainforest growth and atmospheric carbon dynamics linked to interannual temperature variation during 1984–2000. *PNAS* 100 (10):5852–57.

DellaSala, D. 2009. Testimony to the House Subcommittee on National Parks, Forests, and Public Lands for the hearing on "The role of federal lands in combating climate change." March 3, 2009.

Depro, B., B. C. Murray, R. J. Alig, and A. Shanks. 2007. Public land, timber harvest, and climate mitigation: Quantifying carbon sequestration potential on U.S. public timberlands. *Forest Ecology and Management* 255:1122–34.

Feeley, K. J., S. J. Wright, M. N. Nur Spardi, A. R. Kassim, and S. J. Davies. 2007. Decelerating growth in tropical forest trees. *Ecology Letters* 10:461–69.

Fitzgerald, J., D. A. DellaSala, C. May-Tobin. 2009. U.S. House of Representatives Committee on Agriculture biogeographical forum. Unpublished report submitted to the U.S. Congress by the Society for Conservation Biology.

Friedman, T. L. 2008. *Hot, flat, and crowded.* New York: Farrar, Straus and Giroux.

Hansen, J., M. Sato, P. Kharecha, D. Beerling, R. Berner, V. Masson-Delmotte, M. Pagani, M. Raymo, D. L. Royer, and J. C. Zachos. 2008. Target atmospheric CO_2: Where should humanity aim? *Open Atmospheric Science* 2J (2):217–31.

Hanson, D. A., I. Mintzer, J. A. S. Laitner, and J. A. Leonard. 2004. Engines of growth: Energy challenges, opportunities, and uncertainties in the 21st century. Argonne National Laboratory, Illinois.

Harmon, M. E., W. K. Ferrell, and J. F. Franklin. 1990. Effects on carbon storage of conversion of old-growth forests to young forests. *Science* 247:699–702.

———. 2001. Carbon sequestration in forests: Addressing the scale question. *Journal of Forestry* 99:24–29.

Hudiburg, T., B. Law, D. P. Turner, J. Campbell, D. Donato, and M. Duane. 2009. Carbon dynamics of Oregon and northern California forests and potential land-based carbon storage. *Ecological Applications* 19 (1):163–80.

Intergovernmental Panel on Climate Change. 2007. Synthesis report. Geneva, Switzerland. www.ipcc.ch/contact/contact.htm.

Keith, H., B. G. Mackey, and D. B. Lindenmayer. 2009. Re-evaluation of forest biomass carbon stocks and lessons from the world's most carbon-dense forests. *Proceedings of the National Academy of Sciences* 106 (28):11635–40.

Law, B. E., D. Turner, J. Campbell, O. J. Sun, S. Van Tuyl, W. D. Ritts, and W. B. Cohen. 2004. Disturbance and climate effects on carbon stocks and fluxes across Western Oregon, USA. *Global Change Biology* 10:1429–44.

Lichatowich, J. A. 1999. *Salmon without rivers: A history of the Pacific salmon crisis.* Washington, DC: Island Press.

Luyssaert, S., E. Detlef Schulze, A. Börner, A. Knohl, D. Hessenmöller, B. E. Law, P. Ciais, and J. Grace. 2008. Old-growth forests as global carbon sinks. *Nature* 455:213–15.

McNeely, J. 2009. Applying the diversity of international conventions to address the challenges of climate change. *Michigan State Journal of International Law,* Vol. 17, No. 1:123–38.

Mote, P. W., A. F. Hamlet, M. P. Clark, and D. P. Lettenmaier. 2005. Declining mountain snowpack in western North America. *American Meteorological Society* January 2005:39–49.

Nordhaus, W. D., and E. C. Kokkelenberg. 1999. *Nature's numbers: Expanding the national economic accounts to include the environment.* Washington, DC: National Academy Press.

Noss, R. F. 2001. Beyond Kyoto: Forest Management in a time of rapid climate change. *Conservation Biology* 15 (3):578–90.

O'Brien, K., et al. 2009. Rethinking social contracts: Building resilience in a changing climate. *Ecology and Society* 14:12. www.ecologyandsociety.org.

Paine, R. T., M. J. Tegner, and E. A. Johnson. 1998. Compounded perturbations yield ecological surprises. *Ecological Applications* 1998 (1):535–45.

Phillips, O. L., E. O. Luiz, C. Aragão, S. L. Lewis, J. B. Fisher, J. Lloyd, G. López-González, et al. 2009. Drought sensitivity of the Amazon rainforest. *Science* 323:1344–47.

Reusch T. B., A. Ehlers, H. Hammerli, and B. Worm. 2005. Ecosystem recovery after climatic extremes enhanced by genotypic diversity. *Proceedings of the National Academy of Sciences USA* 102:2826–31.

Smithwick, E. A. H., M. E. Harmon, S. M. Remillard, S. A. Acker, and J. F. Franklin. 2002. Potential upper bounds of carbon stores in forests of the Pacific Northwest. *Ecological Applications* 12:1303–17.

Society for Conservation Biology (SCB). 2009. Recommendations for actions by the Obama administration and the Congress to advance the scientific foundation for conserving biological diversity. November 17, 2009. www.consbio.org

Suttle, K. B., M. A. Thomsen, and M. E. Power. 2007. Species interactions reverse grassland responses to climate change. *Science* 315:640–42.

Thompson, I., B. Mackey, S. McNulty, and A. Mosseler. 2009. Forest resilience, biodiversity, and climate change. A synthesis of the biodiversity/resilience/stability relationship in forest ecosystems. Secretariat of the Convention on Biological Diversity, Montreal. Technical Series No. 43.

Turner, D. P., W. D. Ritts, B. E. Law, W. B. Cohen, Z. Yang, T. Hudiburg, J. L. Campbell, and M. Duane. 2007. Scaling net ecosystem production and net biome production over a heterogeneous region in the western United States. *Biogeoscience* 4:597–612.

van Mantgem, P. J., N. L. Stephenson, J. C. Byrne, L. D. Daniels, J. F. Franklin, P. Z. Fulé, M. E. Harmon, et al. 2009. Widespread increase of tree mortality rates in the western United States. *Science* 323:521–24.

Westerling, A. L., H. G. Hidalgo, D. R. Cayan, and T. W. Swetnam. 2006. Warming and earlier spring increase western U.S. forest wildfire activity. *Science* 313:940–43.

Contributors

Paul Alaback, Ph.D., is the Champion Professor of Forest Ecology at the University of Montana. Dr. Alaback has studied the ecology of temperate rainforests throughout the Americas for 30 years, focusing on how climate controls the function and diversity of these forests, as well as how forests respond to disturbances. He currently teaches and conducts research on forest and grassland ecology as well as natural history education.

Ketevan Batsatsashvili, Ph.D., has been assistant professor at Ilia State University since 2008 and a scientist at Georgia's Tbilisi Botanical Garden and Institute of Botany since 2003. She is author or co-author of five scientific publications on the plant diversity of Georgia and served as regional coordinator assistant on the Red List Assessment of Caucasus endemic plants for the IUCN—The World Conservation Union (2006–2009).

Robert P. Cameron is an ecologist and lichenologist responsible for research programs in the Protected Areas Branch of the Nova Scotia Department of Environment and Labour, Canada. Over the past decade, he has published numerous studies on the effects of forestry practices, air pollution, and other stressors on Nova Scotia lichens. This work has focused especially on cyanolichens of coastal forests, including the globally threatened boreal felt lichen, *Erioderma pedicellatum*, and its relative *E. mollissimum*.

Stephen R. Clayden, Ph.D., heads the Botany and Mycology Section at the New Brunswick Museum, Canada. He is a lichenologist and botanist with more than 30 years of experience in field research, conservation, and public education in eastern Canada. His contributions to public education and to developing the herbarium of the New Brunswick Museum as a regional centre for biodiversity research have earned awards from the Canadian Council on Ecological Areas and the Canadian Museums Association.

Frank Lance Craighead, Ph.D., is executive director of the Craighead Environmental Research Institute. His research interests are focal or umbrella species, population and metapopulation persistence, gene flow, habitat connectivity, core and protected areas, and carnivore ecology. He is a member of the IUCN World Committee on Protected Areas, Society for Conservation Biology, Society for Conservation GIS, and The Wildlife Society. He currently serves on the Montana Department of Fish, Wildlife, and Parks Connectivity Working Group and directs the Craighead Center for Landscape Conservation.

Dominick A. DellaSala, Ph.D., is president and chief scientist of the Geos Institute in Ashland, Oregon, and president of the Society for Conservation Biology, North America Section. He is the author of over 150 technical papers and book chapters. His policy work has been featured in *National Geographic*, *Science Digest*, *Science*, *Audubon*, *High Country News*, the *New York Times*, and the *Los Angeles Times*, plus numerous television, radio, and documentary films. These efforts have helped protect millions of hectares of forests in the Pacific Northwest and elsewhere, for which he received leadership awards from the World Wildlife Fund (2000, 2004) and the Wilburforce Foundation (2006).

John M. Fitzgerald is policy director for the Society for Conservation Biology. Fitzgerald is a member of the District of Columbia Bar and has worked with conservation groups, think tanks, whistleblowers, and government officials to secure improvements in federal and international law and policy. His areas of focus are renewable energy, natural resource conservation, anti-corruption measures, international development, and socially and environmentally responsible investing.

Trevor Goward is an internationally recognized authority on the ecology and taxonomy of lichens, with more than 70 technical publications and two books. For more than 25 years he has worked through his consulting firm, Enlichened Consulting Ltd., to provide expertise on lichen systematics and environmental indicators. Since 1988 he has served as curator of lichens at the University of British Columbia. His upcoming book, *Ways of Enlichenment: Macrolichens of Northwest North America*, will be a landmark publication in lichenology.

R. Edward Grumbine, Ph.D., chairs the Masters in Environmental Studies Program at Prescott College in Arizona. He has worked on biodiversity protection on federal lands in the U.S. for over 25 years and is known for contributing to the theory and practice of ecosystem management and large-scale wildlands protection. Grumbine is the author of *Ghost Bears: Exploring the Biodiversity Crisis* and editor of *Environmental Policy and Biodiversity*. His upcoming book is *Where the Dragon Meets the Angry River: Nature and Power in the People's Republic of China*.

Håkon Holien, Ph.D., is an associate professor at the Nord-Trøndelag University College in Steinkjer, Norway. He is the author of 40 scientific papers and two books on lichen ecology, floristics, biogeography, and taxonomy. Since 1980 his main interest has been the boreal rainforests of coastal Central Norway and its particular lichen biogeographical element. He is a member of the group of lichen experts appointed by the Norwegian Biodiversity Information Centre to prepare the national Red List of lichens.

Paul Hosten, Ph.D., has worked with universities, conservation organizations, and federal agencies in southwest Oregon, studying grass, shrub, woodland, and forested wildlands, with a focus on historic condition, distribution, and temporal patterns of change at the stand and landscape scales, and in South Africa at the University of Port Elizabeth, Rhodes University, and the University of Natal. Hosten is currently the terrestrial ecologist at Kalaupapa National Historic Park in Hawaii.

James B. Kirkpatrick, Ph.D., is professor of geography and environmental studies at Australia's University of Tasmania in Hobart, Tasmania. He is author or co-author of over 200 scientific papers and 20 books, primarily on nature conservation. He has conducted field work in the forests of Tasmania, California, Fiji, New Zealand, Victoria, and Northern Territory. He received the Eureka Prize for Environmental Research and an Order of Australia award for his work on forest conservation.

Pavel V. Krestov, Ph.D., directs the laboratory of geobotany programs at the Institute of Biology and Soil Science in Vladivostok, Russia. The rainforests of the Russian Far East were his first love, and his 1996 Ph.D. dissertation was devoted to the last remaining undisturbed Korean pine forest of the middle Sikhote-Alin. He has authored and co-authored over 50 peer-reviewed papers and book chapters published by Opulus Press, Kluwer Academic Publishers, Blackwell Publishing, and Elsevier.

Federico Luebert has worked as an assistant professor of botany at the faculty of forest sciences of the University of Chile (1999–2006) and as a visiting scholar at the Smithsonian Institution (2006). Luebert has contributed to projects on vegetation ecology, plant systematics, and biodiversity conservation, in collaboration with the World Wildlife Fund, Nature Conservancy, NatureServe, and the Chilean environmental agency, CONAMA. He is currently completing his Ph.D. at the Freie Universität in Berlin, supported by the German Academic Exchange Service (DAAD).

John W. McCarthy, Ph.D., is a Jesuit priest and chaplain of St. Mark's College, University of British Columbia, and a forest ecologist, soils scientist, and lichenologist. His work includes examining the structure and dynamics of old-growth boreal forests in western Newfoundland. For his conservation achievements, he received the 2002 Canadian Environment Award Gold Medal. He co-chairs the province's Wilderness and Ecological Reserves Advisory Council, and has written extensively on the spiritual dimensions of environmental ethics and natural areas conservation.

Jeffrey A. McNeely has served as senior science advisor at the IUCN since 1980, overseeing the world's largest conservation network. He has written or edited over 40 books and 500 popular and technical articles, and served on the editorial boards of 14 international journals. He is currently working to link biodiversity to sustainable agriculture, human health, biotechnology, climate change, and energy.

Faisal Moola, Ph.D., is director of the Terrestrial Conservation and Science Program at the David Suzuki Foundation. He has published widely in scientific journals on topics of endangered species, ecology, conservation biology, and environmental policy, and is adjunct professor at the faculty of forestry at the University of Toronto. Moola contributed to significant conservation achievements, including protection of two million hectares of temperate rainforest in British Columbia and the development of sustainability standards for the Forest Stewardship Council.

Yukito Nakamura, Ph.D., is the academic supervisor of the graduate school of forestry at Tokyo University of Agriculture and a member of the Environmental Assessment Committee and Wildlife Management Committee for Kanagawa prefecture. Nakamura is an internationally renowned author, with over 120 technical papers on plant ecology, vegetation science, and landscape ecology.

George Nakhutsrishvili, Ph.D., is principal scientist and chair of the Scientific Council of the Tbilisi Botanical Garden and Institute of Botany, and full professor of life sciences at Ilia State University in Georgia. He also serves as a corresponding member of the Georgian and Austrian Academies of Sciences, and has authored more than 200 scientific and other publications, focusing on the botanical diversity of the Caucasus. He is president of the Georgian Botanical Society, chief of the Biodiversity and Ecological Problems Commission of the Georgian Academy of Sciences, and focal point of the Caucasus Plant Red List Authority at the IUCN Species Survival Commission.

Richard S. Nauman is the conservation scientist and GIS program manager at the Geos Institute in Ashland, Oregon. He has worked on the management and conservation of temperate forests in the northwestern United States for over 20 years, and is a recognized expert in the taxonomy and ecology of terrestrial salamanders in the Klamath-Siskiyou. Nauman is currently writing a book on the national parks and protected areas of Mexico's Baja Peninsula.

Dina I. Nazimova, Ph.D., graduated from Moscow State University in 1959 and works as professor of biology at the Institute of Forest Science in Krasnoyarsk, Russia. She is author or co-author of over 120 papers published in scientific journals and co-author of five monographs. She is a recognized expert in biogeography, forest ecology, and forest typology, with a major focus on the forests of Siberia.

Reed F. Noss, Ph.D., is the Davis-Shine Professor of Conservation Biology at the University of Central Florida. He is president of the Florida Institute for Conservation Science and an international consultant and lecturer. Dr. Noss is the author of more than 260 scientific articles and several books, and served as editor-in-chief of *Conservation Biology* (1993–1997). He is a certified senior ecologist with the Ecological Society of America and an elected fellow of the American Association for the Advancement of Science.

Paul Paquet, Ph.D., is a senior scientist with the Raincoast Conservation Foundation, international consultant, and adjunct professor at the University of Calgary and the University of Manitoba. He is an internationally recognized authority on mammalian carnivores, especially wolves, and was one of the architects of the Large Carnivore Initiative for Europe. His current research focuses on conservation of large carnivores and the effects of human activities on their survival.

Patricio Pliscoff is a researcher with the Institute of Ecology and Biodiversity and RIDES (Research and Resources for Sustainable Development), and has worked as a consultant for conservation agencies such as the World Wildlife Fund, Nature Conservancy, and the Chilean environmental agency, CONAMA. His research is focused on biogeography, conservation biology, and the vegetation and bioclimatology of Chile.

Katrin Ronnenberg is a botanist specializing in dryland research, with a focus on population ecology, vegetation science, and restoration ecology in Mongolia. She has extensive experience in applied aspects of nature conservation of Scottish woodlands. Ronnenberg has surveyed woodlands in western Scotland for the Native Woodland Survey of Scotland, focusing on vegetation, flora, and human impacts.

Jeannine Rossa is a freshwater conservation biologist with over 20 years' experience studying the relationship between forest management and stream ecosystems in the western United States, working with federal land management agencies, universities, and nongovernmental organizations. Her recent work promotes conservation of ecosystem pathways in land-use planning across the wildland, urban, and agricultural sectors in several countries, including South Africa.

John Schoen, Ph.D., is senior scientist for Audubon Alaska in Anchorage and has been involved in Alaskan wildlife research and conservation since 1976, including studying brown bears, black-tailed deer, and mountain goats. He is affiliate professor of wildlife biology at the University of Alaska Fairbanks and has published over 50 scientific and popular articles. Schoen currently serves on the Tongass Futures Roundtable, seeking collaborative solutions for conservation and management of the Tongass, and has awards for conservation leadership from the Alaska Conservation Foundation and the Wilburforce Foundation.

Toby Spribille is a botanist and ecologist specializing in lichens of the North Pacific Rim. As part of his biodiversity research, he has studied and collected species across the boreal forests of the Northern Hemisphere. He is the author of more than 40 technical papers and has identified twenty new species for science. He splits his time between northwest North America and Europe, and is currently based at the University of Graz in Austria.

Nikolai V. Stepanov, Ph.D., heads the herbarium at Krasnoyarsk National University in Krasnoyarsk, Russia. He is author or co-author of 120 publications, including five monographs on the flora of southern Siberia, particularly its historical formation and phytogeographical connections.

David Tecklin has worked on environmental issues in Chile since 1997 and founded the World Wildlife Fund's Chile program in 2002. His research focuses on conservation planning, land conservation initiatives on public and private lands, community-based conservation, and strategies to address threats to biodiversity.

Henrik von Wehrden, Ph.D., is currently working for the Institute of Geobotany in Halle, Germany, on an interdisciplinary project to establish a protection framework for the Mongolian wild ass and specializes in drylands with a focus on southern Mongolia. As a trained geographer with a background in vegetation science, Wehrden combines spatial data and statistical analysis to derive key data for nature conservation.

Nugzar Zazanashvili, Ph.D., has been associate professor at Ilia State University since 2008 and has worked at Georgia's Tbilisi Botanical Garden and Institute of Botany since 1978. He is author or co-author of 57 scientific and 35 popular publications on the landscapes and biodiversity of the Caucasus, and helped develop the Map of Natural Vegetation of Europe (1983–2001). As conservation director for World Wildlife Fund's Caucasus Programme, he coordinated the Caucasus Ecoregional Conservation Plan (2002–2006).

Index

About the Geos Institute

The Geos Institute (www.geosinstitute.org) is a nonprofit, science-based organization dedicated to helping both human and natural communities prepare for a changing climate. To this end, the Geos Institute applies the best available science to natural-resource-conservation issues through its scientific publications and its ability to link respected scientists to decision makers. The Geos Institute serves communities and land managers by assisting them with on-the-ground planning for climate change through the emerging field of climate change readiness ("adaptation") and efforts to reduce greenhouse-gas pollutants ("mitigation").